The Fibonacci Sequence
and
Beyond

by
Bruce R. Gilson

2012

Table of Contents

List of Tables

Preface.

The thirteenth-century mathematician Leonardo of Pisa (sometimes rendered in Italian as Leonardo Pisano) was one of the few people in that era to contribute to the progress of mathematics in Europe. He wrote a book, whose title is usually rendered *Liber Abaci*, but which I have read should correctly be rendered as *Liber Abbaci*, about methods of calculation. In this book a problem was posed, ostensibly about the reproduction of rabbits, but really in order to propose the sequence mentioned in the title of this book. Leonardo was also known as *filius Bonacci, i. e.* the son of Bonaccio, and in honor of the shortened form *Fibonacci* of that name the sequence has been named the *Fibonacci sequence*.

For a long time the interesting properties of this sequence were forgotten. In the nineteenth century, a French mathematician, Édouard Lucas, revived the interest in this sequence, and at the same time introduced a closely related sequence, which has been named in his honor the *Lucas sequence*.

This book is about the well-known Fibonacci sequence, but I have attempted to do a little more than most books about this sequence generally do. I have covered other sequences, including the Lucas sequence and geometric sequences, as well as some less-known types, which I try to cover in a way to illustrate relationships between them. (As stated in Chapter 3, almost all the sequences discussed in this book will be *recurrent sequences*, as defined in that chapter, so it could have been entitled "The Fibonacci Sequence and Other Recurrent Sequences." However, the term "recurrent sequence" is not as well-known as the name of the Fibonacci sequence, so that title was not chosen.) I hope that this book will be a useful contribution to the literature. (Two concepts that I wish to emphasize, that I believe I am introducing in this book as I have no recollection of seeing anywhere else, are those of *co-recurrent sequences* and the *prototypic co-recurrent sequence*, defined in Chapter 3. In addition, many of the interesting properties of the Fibonacci sequence, which are often introduced without reference to the fact that they are special cases of recurrent sequence properties, would be easier to understand if they were put into this more general context. This includes the expression of the terms of the Fibonacci sequence as powers of τ and τ', often introduced as an odd fact, unrelated to the whole idea of recurrent sequences, though it follows quite directly from the fact that the Fibonacci sequence is a 3-term recurrent sequence, and all n-term recurrent sequences can be expressed as $n\text{-}1$-fold geometric sequences, as shown in Chapter 4.)

One more point of terminology: Some books refer to "Fibonacci numbers" and "Lucas numbers." I have tried to avoid this terminology (which is even in the *titles* of books on the subject: see the Bibliography!) because this is not a property of the *individual numbers* (after all, some numbers, such as 3, are in *both* sets!) but really of the *sequences*. It is really meaningless to try to base anything on the fact that "3 is a Fibonacci number" or "3 is a Lucas number," but taking the fact that "1, 1, 2, 3" are *4 consecutive terms of the Fibonacci sequence* enables us to say many things about this collection.

The only prerequisite for this book is the algebra that most of us have learned in high school (or these days even earlier), but it does presuppose a willingness to look at, and follow, the manipulation of algebraic equations. Those who hate mathematics probably have not even opened the book, so I assume that readers are at least willing to read books with a certain amount of mathematics; I believe that if you spend the time to look at the ideas inside, you will see a lot of nice patterns, which to me is the beauty of mathematics.

Obviously, every author who writes on this subject has favorite topics to consider, and there are a number of excellent books available which I do not wish to damn with faint praise. The Bibliography lists other works that include material that I have not chosen to cover, but which the reader may find interesting as well. If this book arouses your interest, you may want to read these other books as well; please be aware that they may use different notations from this book, but they should be understandable anyway.

Because I have not had the luxury of getting comments before publication, or even the opportunity (or obligation!) to submit the work for editing by another person, even more than most authors I must apologize that "all errors are mine and mine alone." I hope there will not be many. If anybody finds errors in this book, or has any suggestions about things that should have been covered, please send an e-mail to me at brg1942@gmail.com. I guarantee that any corrections or suggestions for improvement would be appreciated.

Bruce R. Gilson

November 15, 2012

Chapter 1: Some preliminary definitions.

Terms that are defined in this book will be indicated in ***bold italics like this***.

A ***sequence*** or ***progression*** is a set of numbers arranged in order. Both words should be understood as equivalent; the reason for using both these terms is simply common usage: generally the term "geometric progression" is used rather than "geometric sequence," but "Fibonacci sequence" rather than "Fibonacci progression." The individual numbers in a sequence are called its ***terms***; they do not need to be all different, but their order in the sequence is important.

Normally, a sequence will be designated by a capital letter, and the individual terms by the same letter with a subscript. So if a sequence is denoted by A, its first term is A_1, its second as A_2, etc.

Although it is normal to define a first, second, etc. term of a sequence, when there is a specific recipe for constructing the terms, and the rule gives meaningful results for subscript 0, –1, –2, etc., these will be considered as meaningful values of A_0, A_{-1}, A_{-2}, etc.

The ***Fibonacci sequence***, symbolized F, is defined by $F_1 = F_2 = 1, F_n = F_{n-1} + F_{n-2}$. The ***Lucas sequence***, symbolized L, is defined by $L_0 = 2, L_2 = 1, L_n = L_{n-1} + L_{n-2}$. A ***generalized Fibonacci sequence***, symbolized G, is defined by $G_n = G_{n-1} + G_{n-2}$, where G_1 and G_2 can be *any* numbers. (It should be noted that any generalized Fibonacci sequence with $G_1 = G_2$ is simply the Fibonacci sequence with all its terms multiplied by a constant, while $G_1 = 1, G_2 = 2$ simply generates a sequence with $G_n = F_{n-1}$. So most combinations of small numbers for starting terms, other than the two defining the Fibonacci and Lucas sequences, do not yield anything new.) Many of the special properties of the Fibonacci sequence that one sees in books are really traceable to its being an instance of the generalized Fibonacci sequence, and in this book will be derived for generalized Fibonacci sequences rather than just for the Fibonacci sequence, with the special form that they take for the Fibonacci sequence being obtained by simply putting $F_1 = F_2 = 1$ in the generalized Fibonacci sequence formula. In other cases, the formula will be derived in a way that is even more general, based on the fact that a generalized Fibonacci sequence is a special case of a recurrent sequence (defined in Chapter 3), and so the formula can be derived for recurrent sequences, specialized for generalized Fibonacci sequences, and finally specialized even further for the Fibonacci sequence.

Sequences can be defined in two different ways. A formula can be given, containing the term number n, which can be used to calculate each term of the sequence (an ***explicit definition*** of the sequence), or some few terms of the sequence can be assigned values and a formula for computing other terms from given terms can be given, as was done for the Fibonacci and Lucas sequences in the preceding paragraph (a **recursive definition** of the sequence). Some of the discussions in this book will involve producing equivalent recursive definitions when sequences are explicitly defined or equivalent explicit definitions when they are recursively defined. (For example, a geometric progression P can be recursively defined by $P_1 = a$, $P_{n+1} = rP_n$, or explicitly by $P_n = ar^{n-1}$.)

The next few definitions, which refer to quotients, remainders, and other related concepts, are not going to be used until much later in the book, but they are given here simply because it is desirable to have all the fundamental definitions of terms together in one place. The reader is probably, from elementary arithmetic, familiar with the process of integer division to give a *quotient* and *remainder*. Formally, when one considers the division of one integer n (the **dividend**) by another integer m (the **divisor**), the **quotient** and **remainder** are the two integers q and r such that $n = qm + r$, choosing q to be the integer that gives $0 \le r < m$. This definition is universally used when $n > 0$; some computer programming languages, at least, use a different criterion when $n < 0$ (namely that q is chosen to be the integer that gives $-m < r \le 0$), but in this book, as in most mathematical work, it will be universally taken to give $0 \le r < m$. In this book it will not be necessary to deal with negative values of the divisor m, and since division by 0 is never permitted, and division by 1 never leaves a remainder, when talking about remainders it will always be assumed that $m \ge 2$. Frequently it is useful to consider the set of all integers that give the same remainder when divided by a particular divisor; this is called a **residue class** (labeled with the *remainder* r and the divisor, which in this case is called a **modulus,** m). So the residue class 3 under the modulus 5 is the set $\{..., -22, -17, -12, -7, -2, 3, 8, 13, 18, ...\}$, for example. If two integers n_1 and n_2 are in the *same* residue class under a given modulus, they are said to be **congruent** under that modulus, and the symbol $\equiv$ is used: writing, for example,

$$18 \equiv 3 \ (\text{mod } 5).$$

(Note that it is not necessary that both, or even one, be a simple integer; the 18 and 3 above could be replaced by any algebraic expressions.) An equivalent definition of congruence is sometimes met with: some books define $n_1 \equiv n_2$ (mod m) as meaning that $n_1 - n_2 = km$ for some integer k; however this is in fact simply another way of stating the same thing. If $n \equiv 0$ (mod m), which means that $n = km$ for some integer k, one says that n is a **multiple** of m, that n is **divisible** by m, or that m **divides** n **exactly** (sometimes leaving out the word "exactly"), all of which mean the same thing. Later chapters will discuss residue classes, congruence, and divisibility, but they will not be of prime concern in the next few chapters of this book; as stated above, these definitions were included at this point only to keep all the fundamental definitions together in one chapter.

Chapter 2: Some properties of generalized Fibonacci sequences.

It might be noted that, while it is usual to define a generalized Fibonacci sequence by a recursive definition where G_1 and G_2 (or G_0 and G_1) are given as well as the formula $G_n = G_{n-1} + G_{n-2}$ (which will be termed the **generalized Fibonacci sequence definition formula**) from the previous chapter, any two consecutive terms can be specified, and that formula used. To make it easier to extend the series backward, if two rather later terms are supplied, the generalized Fibonacci sequence definition formula can be rewritten as $G_{n-2} = G_n - G_{n-1}$. And of course, the sequence can be extended backward using this formula to produce values for G_0, G_{-1}, G_{-2}, etc., as implied in the previous chapter.

In particular, it can be seen that if there are any two generalized Fibonacci sequences, G and G', such that $G'_p = G_q$ and $G'_{p+1} = G_{q+1}$ for some p and q, then $G'_{p+t} = G_{q+t}$ for all t. (Just apply the generalized Fibonacci sequence definition formula as many times as necessary.) We will describe a generalized Fibonacci sequence G' as a **shifted form** of another generalized Fibonacci sequence G whenever they have this kind of relationship. (The difference $|q - p|$ will be termed the **amount of shift**.) Shifted forms of the Fibonacci and Lucas sequences will be referred to more concisely as **shifted Fibonacci sequences** and **shifted Lucas sequences**, respectively.

Again, by applying the generalized Fibonacci sequence definition formula as many times as necessary, it is clear as well that if there are any two generalized Fibonacci sequences, G and G', and a third sequence G'' is created with $G''_i = G_i + G'_i$ for all i, the sequence G'' will be a generalized Fibonacci sequence as well. The two sequences can be related by a shift or totally unrelated. I will have occasion to use this result at various points in this book.

Another notable property is that if any two consecutive terms of a generalized Fibonacci sequence have a common factor, then *all* terms in that sequence share that common factor. For if

$$G_i = cx_1 \text{ and}$$

$$G_{i+1} = cx_2,$$

where c, x_1, and x_2 are integers, then

$$G_{i-1} = G_{i+1} - G_i$$

$$= cx_2 - cx_1$$

$$= c(x_2 - x_1), \text{ and}$$

$$G_{i+2} = G_{i+1} + G_i$$

$$= cx_2 + cx_1$$

$$= c(x_2 + x_1),$$

and this process can be continued indefinitely. In particular, it can be seen that no generalized Fibonacci sequence that has a 1 as any term can have two consecutive terms that share a common factor, and thus, specifically, it would be impossible for any two consecutive terms of either the Fibonacci or the Lucas sequence to have any common factors, because both sequences contain 1's.

Chapter 3: Recurrent sequences.

The generalized Fibonacci sequence definition formula is a special case of a *recursion formula*. In general, if a sequence has a definition of the form $A_n = k_1A_{n-1} + k_2A_{n-2} + \ldots + k_pA_{n-p}$, the sequence is termed a **recurrent sequence**, and the formula defining it is called a **recursion formula.** (Note that the term "recursive definition" was defined in a more general manner than this, where *any* form of definition in terms of previous terms is included, in Chapter 1. This difference between "recursion formula (for terms in a sequence)" and "recursive definition (of a sequence)" will be maintained in this book. And not every sequence defined by a *recursive definition* is a *recurrent sequence* by this definition, though most of them are.) The coefficients k_1, k_2, ..., k_n are termed the **recursion coefficients,** and it will occasionally be useful to refer to the **first recursion coefficient** k_1 or the **second recursion coefficient** k_2.

The concept of "two recurrent sequences with the *same recursion formula*" turns out to be quite important, yet I do not recall ever seeing it given a name anywhere, although in this book it will often be useful to talk about pairs of sequences related this way. So they will be referred to as **co-recurrent sequences.** (With this definition, a generalized Fibonacci sequence could simply be defined as any sequence co-recurrent with the Fibonacci sequence.) As noted, by this definition, any generalized Fibonacci sequence is a recurrent sequence, but so are a lot of other types of sequences. For example, if A is a geometric progression P, with $P_i = ar^{i-1}$. then setting $A_1 = a$, $p = 1$, $k_1 = r$ will produce that sequence. And similarly, for an arithmetic progression, by writing it in the form $A_i = A_{i-1} + (A_{i-1} - A_{i-2}) = 2A_{i-1} - A_{i-2}$, it can be seen that it can also be put in the form of a recurrent sequence, with $A_1 = a$, $p = 2$, $k_1 = 2$, $k_2 = -1$.

One important characteristic of a recurrent sequence is the value of p, though it is usually $p+1$ that is used to characterize them. Writing the recursion formula in the form $A_n - k_1A_{n-1} - k_2A_{n-2} - \ldots - k_pA_{n-p} = 0$, which can be called the **zero-right-hand-side form**, the left-hand side of the equation has $p+1$ terms, so the relationship is called a **$p+1$-term recurrence relation**. So the generalized Fibonacci sequence definition formula is a 3-term recurrence relation, as is the formula for an arithmetic progression given above, while the formula for a geometric progression is a 2-term recurrence relation. In general, "recurrent sequence with an n-term recurrence relation" will be shortened to "n-term recurrent sequence."

Most of the time, in this book, 3-term recurrence relations will be discussed, so in general, after this chapter, the term "recurrent sequence" can be understood to mean "3-term recurrent sequence."

They can be thought of as further generalizations of generalized Fibonacci sequences, where k_1 and k_2 need not be 1.

Almost all the sequences discussed in this book will be recurrent sequences, so it could have been entitled "The Fibonacci Sequence and Other Recurrent Sequences." However, the term "recurrent sequence" is not as well-known as the name of the Fibonacci sequence, so that title was not chosen.

In the preceding chapter, it was stated that if there are any two generalized Fibonacci sequences, G and G', such that $G'_p = G_q$ and $G'_{p+1} = G_{q+1}$ for some p and q, then $G'_{p+t} = G_{q+t}$ for all t. It can be seen that by the same reasoning, if there are any two co-recurrent sequences, A and A', such that $A'_p = A_q$ and $A'_{p+1} = A_{q+1}$ for some p and q, then $A'_{p+t} = A_{q+t}$ for all t. As was done with generalized Fibonacci sequences, a recurrent sequence A' will be described as a **shifted form** of another recurrent sequence G whenever they have this kind of relationship, and the difference $|q - p|$ will be termed the **amount of shift**. It is clear that if one recurrent sequence, A, is related to another recurrent sequence, A', in such a way that there exists some p and q such that $A'_{p+t} = A_{q+t}$ for all t, then one is a shifted form of the other, and the two are co-recurrent, though co-recurrent sequences are not necessarily related by shifting.

Besides shifting, it will frequently be found useful, when starting with a given recurrent sequence, to derive various sequences which are co-recurrent with it. Just as the term "co-recurrent sequence" (and later, the term "difference-of-products formula") are being introduced in this book because I know of no existing term in current terminology for these concepts, I will introduce some new terminology for certain types of co-recurrent sequences.

First of all, given any recurrent sequence, it will be useful to define another recurrent sequence, co-recurrent with it, that occurs in a number of formulas which will be given in this book. The name **"prototypic co-recurrent sequence"** will be used in this book with the following significance: For any recurrent sequence A which is defined by the recursion formula $A_i = k_1 A_{i-1} + k_2 A_{i-2}$, construct a sequence C with $C_i = k_1 C_{i-1} + k_2 C_{i-2}$ such that $C_0 = 1$ and $C_1 = k_1$. This will be termed the **prototypic co-recurrent sequence** of A. It should be noted that any two co-recurrent sequences have the *same* prototypic co-recurrent sequence. For generalized Fibonacci sequences, the prototypic co-recurrent sequence A is the Fibonacci sequence (but shifted by 1: $A_j = F_{j+1}$).

Another concept that will be required later is that which will be termed, in this book, the *standard **co-recurrent basis sequences**.* For any recurrent sequence A, if two co-recurrent sequences A' and A'' are constructed, with

$$A'_0 = 1, A'_1 = 0, A''_0 = 0, \text{ and } A''_1 = 1,$$

and A'_n and A''_n are assumed to be computed from the same recurrence relation as A_n, then A' and A'' will be termed the two **standard co-recurrent basis sequences** of A. (Occasionally, the term "first standard co-recurrent basis sequence" is used for A', and "second standard co-recurrent basis sequence" for A''.) Again, as with the prototypic co-recurrent sequence, all co-recurrent sequences share a pair of standard co-recurrent basis sequences. (Note that in fact, the first standard co-recurrent basis sequence A' is not quite the same as the prototypic co-recurrent sequence defined earlier, but differs by one in the subscripts of each term; A'_n here is what was symbolized as C_{n-1} in the previous paragraph.) The standard co-recurrent basis sequences will be used later in the procedure to generate some useful recurrent sequence formulas. I might be criticized here for introducing the term "standard co-recurrent basis sequences" here, because this is hardly a *standard* concept in the literature; it is being introduced in this book. However, *within this book* this pair of co-recurrent basis sequences is a standard way to resolve a recurrent sequence into a sum of two, and it should be, at least unless the concept becomes adopted generally, understood as a standard within the book.

It might be questioned that, in this book, it has been determined desirable to introduce *both* the concept of the standard co-recurrent basis sequences and that of the prototypic co-recurrent sequence, when the latter is just the first standard co-recurrent basis sequence shifted by 1. There is a justification for this, however: when resolving a recurrent sequence into the sum of multiples of the two standard co-recurrent basis sequences, it turns out that the first one in the form I use it gives a simpler formula than if the prototypic co-recurrent sequence had been chosen instead, while in other formulas, in which the prototypic co-recurrent sequence appears, choosing the first standard co-recurrent basis sequence instead would unduly complicate the formula. So adopting both makes both sets of formulas appear in the simplest form.

Chapter 4: Multiple geometric sequences.

The term "multiple geometric sequence" is not in common usage, but it is desirable to have a term to describe such sequences, and so it will be defined now. A *multiple geometric sequence* is a sequence obtained by the term-by-term addition of two or more geometric progressions. Specifically, if P, P', P'', P''', etc. are geometric progressions, the sequence A with $A_i = P_i + P'_i$ is a *double geometric sequence,* the sequence A with $A_i = P_i + P'_i + P''_i$ is a *triple geometric sequence,* the sequence A with $A_i = P_i + P'_i + P''_i + P'''_i$ is a *quadruple geometric sequence,* etc.

Multiple geometric sequences can always be expressed as recurrent sequences. In this chapter, this will be shown only for *double* geometric sequences, but the same procedure can be applied for all multiple geometric sequences:

Suppose that the common ratios of P and P' are r and r'. Then:

$$A_i = P_i + P'_i = r^2 P_{i-2} + r'^2 P'_{i-2}$$

$$A_{i-1} = P_{i-1} + P'_{i-1} = r P_{i-2} + r' P'_{i-2}$$

$$A_{i-2} = P_{i-2} + P'_{i-2}.$$

Since a recurrence relation is to be established, set

$$r^2 P_{i-2} + r'^2 P'_{i-2} = k_1(r P_{i-2} + r' P'_{i-2}) + k_2(P_{i-2} + P'_{i-2}).$$

This requires

$$r^2 P_{i-2} + r'^2 P'_{i-2} = (k_1 r P_{i-2} + k_1 r' P'_{i-2}) + (k_2 P_{i-2} + k_2 P'_{i-2})$$

or

$$(r^2 P_{i-2} - k_1 r P_{i-2} - k_2 P_{i-2}) + (r'^2 P'_{i-2} - k_1 r' P'_{i-2} - k_2 P'_{i-2}) = 0$$

$$(r^2 - k_1 r - k_2)P_{i-2} + (r'^2 - k_1 r' - k_2)P'_{i-2} = 0$$

But this will be true for all terms only if $r^2 - k_1 r - k_2 = r'^2 - k_1 r' - k_2 = 0$. So setting these up as simultaneous equations:

$$k_1 r + k_2 = r^2$$

$$k_1 r' + k_2 = r'^2$$

and solving,

$$k_1 (r - r') = r^2 - r'^2,$$

$$k_1 = (r^2 - r'^2)/(r - r') = r + r',$$

$$k_2 = r^2 - k_1 r = r^2 - (r + r')r = r^2 - (r^2 + rr') = -rr'.$$

Since this is true for *all* the terms of the progression, the double geometric sequence is expressed as a 3-term recurrent sequence with the specified recursion coefficients k_1 and k_2. In general, any multiple geometric sequence formed by summing n geometric sequences can be written as an $n+1$-term recurrent sequence. And the formulas above, which accomplish this, will be termed the **double-geometric sequence to recurrent sequence formulas.**

Clearly, the reverse can be done as well. Almost any $n+1$-term recurrent sequence can be written as a multiple geometric sequence formed by summing n geometric sequences by reversing the process (see below for a rare case of an example where it cannot be done). For if $k_1 = r + r'$, $k_2 = -rr'$, one can solve for r and r' in terms of k_1 and k_2. The resulting formulas, which will be termed the **recurrent sequence to double-geometric sequence formulas,** are:

$$r = [k_1 + \sqrt{(k_1^2 + 4k_2)}]/2,$$
$$r' = [k_1 - \sqrt{(k_1^2 + 4k_2)}]/2.$$

(The choice of which sign to use for r and which for r' is, of course, arbitrary.)

One thing that should be noted is that an n-term recurrent sequence is not determined by the recurrence relation alone; it requires $n-1$ of the terms of the sequence to be fixed as well. So, for a 3-term recurrent sequence, two additional items, normally either A_0 and A_1 or A_1 and A_2, are needed to determine the sequence itself. Similarly, in a multiple geometric sequence, the common ratios alone do not determine the terms; at least a number of terms equal to the number of series being summed term-by-term is necessary. Since this is exactly the same number required for the corresponding recurrent sequence, as was shown here for the specific case of a 3-term recurrent sequence corresponding to a double geometric sequence, exactly the same number of items is required to specify a sequence, whether it is conceived of as an $n+1$-term recurrent sequence or an n-fold multiple geometric sequence.

One would think that every arithmetic progression is also a double geometric sequence. For, as was shown in Chapter 3, any arithmetic progression can be written in the form of a recurrent sequence, with $A_1 = a$, $p = 2$, $k_1 = 2$, $k_2 = -1$. But if one tries to apply the recurrent sequence to double-geometric sequence formulas, it gives

$$r = [2 + \sqrt{(4 - 4)}]/2,$$
$$r' = [2 - \sqrt{(4 - 4)}]/2,$$

both of which are 0. So the normal procedure to convert recurrent sequences to double geometric sequences fails in this case. One can see that this is because $k_2 = -k_1^2/4$. So whenever $k_2 = -k_1^2/4$, whatever the values of k_1 and k_2 are, the process fails to generate a double geometric sequence.

An important thing to note is that in a multiple geometric sequence, as one proceeds further along, the progression with the largest-absolute-valued common ratio dominates the sequence. This is true, no matter how close the absolute values of the common ratios are, and no matter how small the coefficient of the common ratio in that progression may be compared to the coefficients in the other progressions summed in the multiple geometric sequence.

To see this, suppose A is a double geometric sequence that can be written

$$A_i = a_1 r^i + a_2 r'^i.$$

Of the two common ratios r and r', determine which of the quantities has the larger absolute value, and call that one r and the other one r'. So it can be assumed that that $|r| > |r'|$; if the reverse is true, just interchange the labels so that the one which had been called r is designated r' and vice versa. With that in mind, the equation can be rewritten

$$A_i = a_1 r^i[1 + x(r'/r)^i],$$

where $x = a_2/a_1$. Since $|r| > |r'|$, it is seen that, when i is very large, $(r'/r)^i$ becomes very small, and thus the factor $1 + x(r'/r)^i$ is close to 1 no matter what the values of x, r, and r' are, so that one can put $A_i \approx a_1 r^i$.

So even in a sequence like

$$A_i = 10^5(2.9)^i + 3^i,$$

for large enough i, A_i will be approximately 3^i. This should be kept in mind when working with multiple geometric sequences, and will come up more than once in subsequent chapters.

The term "multiple geometric sequence" (and its special cases such as "double geometric sequence," "triple geometric sequence," etc. are to my knowledge being introduced for the first time in this book. And no equivalent terms appear anywhere else as far as I am aware. However, I believe the concept is important enough to have a name. While all multiple geometric sequences are also recurrent sequences, and most of this book will deal with them in that form, it will occasionally be useful to consider them as multiple geometric sequences and it should always be kept in mind that one can go back and forth between both representations of the same sequence.

Chapter 5:   The golden ratio.

Following the arguments of Chapter 4, it is clear that since the generalized Fibonacci sequence definition formula is a recurrence relation, every generalized Fibonacci sequence can be put into the form of a double geometric sequence. And in fact, since the coefficients in the generalized Fibonacci sequence definition formula are always 1 and 1, the common ratios of the two geometric sequences summed will be the same for every generalized Fibonacci sequence. This chapter will begin by producing a formula giving the double geometric sequence expression for any specified generalized Fibonacci sequence, which then can be used to give the specific double geometric sequence expressions for the commonly named sequences.

As stated in the preceding paragraph, the coefficients in the generalized Fibonacci sequence definition formula are always 1 and 1. So, substituting $k_1 = 1$ and $k_2 = 1$, in the recurrent sequence to double-geometric sequence formulas of Chapter 4, the **common** ratios of the two double geometric sequences are:

$$r = (1 + \sqrt{5})/2,$$

$$r' = (1 - \sqrt{5})/2.$$

(Note that the second of these is a negative number, since $1 < \sqrt{5}$.) These two numbers are extremely important in the study of Fibonacci and related sequences, and they deserve symbols. The first of these is frequently referred to as the **golden ratio,** or the **ratio of the golden section.** As the word "section" means "cutting" and the Greek word for a cutting is tome⁻ (τομη), the Greek letter τ is used to denote this number (though the Greek letter φ, the initial of the sculptor Phidias in Greek, has also been used; however, it has been established that Phidias never used the golden ratio in his designs; the use of τ in this book agrees with the best books available, Dunlap's and Vajda's books cited in the Bibliography). The second of these numbers is also denoted in some works by φ, and equals both $-1/\tau$ *and* $1 - \tau$, but since there is no universally recognized symbol for this number it will be denoted in this book by τ'. (Among the authors of books cited in the Bibliography at the end of this book, Vajda uses σ, while Dunlap uses φ, for example.)

That $\tau' = 1 - \tau$ and $\tau' = -1/\tau$ can easily be verified; putting them together to produce $1 - \tau = -1/\tau$ and multiplying by τ gives

$$\tau - \tau^2 = -1$$

or, rearranging terms,

$$\tau^2 - \tau - 1 = 0.$$

But since

$$\tau' = 1 - \tau \text{ implies } \tau = 1 - \tau'$$

and

$$\tau' = -1/\tau \text{ implies } \tau = -1/\tau',$$

the exact same reasoning shows that $\tau'^2 - \tau' - 1 = 0$ as well. (Of course, solving $x^2 - x - 1 = 0$ gives as solutions the two numbers here designated as τ and τ', which gives the same result.) In much of what follows, the facts that

$$\tau' = 1 - \tau$$

and

$$\tau' = -1/\tau$$

(sometimes in the form $\tau + \tau' = 1$ and $\tau\tau' = -1$), as well as that

$$\tau^2 = \tau + 1,$$

will be used to simplify many algebraic expressions.

These numbers τ and τ' also have one other interesting property. Unlike any other geometric progressions, those with **common** ratio equal to one of these numbers can be shown to be generalized Fibonacci sequences as well. This can be shown by a backwards argument: suppose a sequence A is both a geometric progression whose common ratio is r and a generalized Fibonacci sequence. Since A is a geometric progression, for any n the equations $A_{n+1} = rA_n$ and $A_{n+2} = r^2A_n$ hold. But since A is a generalized Fibonacci sequence, the equation $A_{n+2} = A_{n+1} + A_n$ also holds. Combining these:

$$A_{n+2} = A_{n+1} + A_n$$
$$r^2A_n = rA_n + A_n$$
$$r^2A_n - rA_n - A_n = 0$$
$$(r^2 - r - 1)A_n = 0,$$

but since this holds for *all n*, this means that

$$r^2 - r - 1 = 0.$$

Because of the relationship of this quadratic equation to generalized Fibonacci sequences in general, and the Fibonacci sequence in particular, the equation has been termed the **Fibonacci quadratic equation**. The solutions to the equation are

$$r = (1 \pm \sqrt{5})/2,$$

that is, the numbers here designated τ and τ'.

The fact that the powers of τ form a generalized Fibonacci sequence can be seen in Table 1, below. If any two consecutive are added (take any two consecutive numbers in the second column) it can be seen that the sum is the next number in the column.

i	τ^i
0	1
1	1.6180340
2	2.6180340
3	4.2360680
4	6.8541020
5	11.0901699
6	17.9442719
7	29.0344419
8	46.9787138
9	76.0131556
10	122.9918694
11	199.0050250
12	321.9968944
13	521.0019194
14	842.9988138
15	1364.0007331
16	2206.9995469
17	3571.0002800
18	5777.9998269
19	9349.0001070
20	15126.9999339
21	24476.0000409
22	39602.9999747
23	64079.0000156
24	103681.9999904
25	167761.0000060
26	271442.9999963
27	439204.0000023
28	710646.9999986

Table 1: Powers of τ.

Many other books have been devoted to the properties of τ, usually discussing τ' as well (though frequently using different symbols from τ and τ'). While most other books I have seen of the

Fibonacci sequence, or on both the Fibonacci and Lucas sequences, devote more space to the discussion of these numbers than this book does (and in particular, see the books in the Bibliography), this does not mean that I think the topic is unimportant, but merely that, because this book treats these sequences in the context of the more general recurrent sequences and double geometric sequences, in which τ and τ' play no specific role, there is less reason to emphasize the importance of τ and τ'. You are especially referred to Livio's book for a detailed discussion of the golden ratio. For the purposes of this book, the most important properties of τ and τ' can be summarized:

1. These quantities relate to each other according to the very symmetric equations $\tau + \tau' = 1$ and $\tau\tau' = -1$, which can also be written as $\tau = 1 - \tau'$, $\tau' = 1 - \tau$, $\tau' = -1/\tau$, and $\tau = -1/\tau'$.

2. Any geometric progression whose common ratio is τ or τ' is also a generalized Fibonacci sequence. Any generalized Fibonacci sequence which is also a geometric progression must have a common ratio which is τ or τ'.

3. Any other generalized Fibonacci sequence is expressible as a double geometric sequence whose common ratios are always τ and τ'.

4. As will be shown in a later chapter, in all generalized Fibonacci sequences, the ratio of consecutive terms gets nearer to τ as the progression proceeds.

5. All powers of τ (and of τ') can be expressed in the form (involving consecutive Fibonacci sequence terms) $F_n + F_{n+1}\tau$.

Chapter 6:   Explicit double geometric sequence formulas for recurrent sequences and generalized Fibonacci sequences.

In this chapter, an ***explicit*** formula is derived for any recurrent sequence that has been expressed by a recursive definition, and the formulas are also specialized to the case of generalized Fibonacci sequences. As noted in Chapter 4, the determination of the two ***common*** ratios r and r' does not fully determine the double geometric sequence which represents a particular recurrent sequence. It is necessary as well to determine the quantities a_1 and a_2 in the equation

$$A_n = a_1 r^n + a_2 r'^{\,n}$$

as well. To do this, start with the ***recurrent*** sequence to double-geometric sequence formulas of Chapter 4:

$$r = [k_1 + (k_1^2 + 4k_2)]/2,$$

$$r' = [k_1 - (k_1^2 + 4k_2)]/2$$

The quantities a_1 and a_2 can be calculated by putting any two terms of the sequence into this formula and solving the resulting equations as simultaneous equations. The two choices that are most common are $i = 0, 1$ or $1, 2$. For the first case, the specific equations

$$A_0 = a_1 r^0 + a_2 r'^{\,0} = a_1 + a_2,$$

$$A_1 = a_1 r^1 + a_2 r'^{\,1} = a_1 r + a_2 r'$$

must be obeyed, where the simplifications are possible since any number to the zero power is 1 and any number to the 1st power is itself. These can be solved simultaneously, giving

$$a_1 = (A_1 - A_0 r')/(r - r'),$$

$$a_2 = (A_0 r - A_1)/(r - r').$$

For the second case, the specific equations

$$A_1 = a_1 r^1 + a_2 r'^{\,1} = a_1 r + a_2 r',$$

$$A_2 = a_1 r^2 + a_2 r'^{\,2}$$

must be obeyed. These can be solved simultaneously, giving

$$a_1 = (A_2 - A_1 r')/(r - r')r,$$

$$a_2 = (A_1 r - A_2)/(r - r')r'.$$

Thus, if one writes either

$$A_n = (A_1 - A_0 r')r^n/(r - r') + (A_0 r - A_1)r'^{\,n}/(r - r'),$$

or

$$A_n = (A_2 - A_1 r')r^{n-1}/(r - r') + (A_1 r - A_2)r'^{n-1}/(r - r'),$$

with

$$r = [k_1 + \sqrt{(k_1^2 + 4k_2)}]/2 \text{ and}$$
$$r' = [k_1 - \sqrt{(k_1^2 + 4k_2)}]/2,$$

an explicit formula for A_n is derived.

These two sets of expressions for a_1 and a_2 look different, but of course they must be equivalent for both to be correct. To demonstrate that they are in fact equivalent, first consider the expressions for a_1. The first set of expressions has

$$a_1 = (A_1 - A_0 r')/(r - r'),$$

Since $rr' = -k_2$, upon multiplying numerator and denominator by r, this becomes

$$a_1 = (A_1 r - A_0 rr')/(r - r')r$$
$$= (A_1 r + k_2 A_0)/(r - r')r;$$

while from the second set of equations, because $A_2 = k_2 A_0 + k_1 A_1$,

$$a_1 = (A_2 - A_1 r')/(r - r')r$$
$$= [(k_2 A_0 + k_1 A_1) - A_1 r']/(r - r')r$$
$$= [k_2 A_0 + (k_1 - r')A_1]/(r - r')r.$$

But again, since $k_1 = r + r'$, one can replace $k_1 - r'$ by r, so

$$a_1 = (k_2 A_0 + r A_1)/(r - r')r,$$

the same expression, except rearranged. Similarly, looking at the first expression for a_2:

$$a_2 = (A_0 r - A_1)/(r - r'),$$

and multiplying numerator and denominator both by r' and using $rr' = -k_2$,

$$a_2 = (A_0 r - A_1)/(r - r')$$
$$= (k_2 A_0 + A_1 r')/(r' - r)r'.$$

The second expression for a_2 is

$$a_2 = (A_1 r - A_2)/(r - r')r'$$
$$= [A_1 r - (k_2 A_0 + k_1 A_1)]/(r - r')r'$$

$$= [(r - k_1)A_1 - k_2 A_0]/(r - r')r'$$

$$= [k_2 A_0 - (r - k_1)A_1]/(r' - r)r'$$

$$= (k_2 A_0 - r'A_1)/(r' - r)r'.$$

Again, this can be seen to be the same equation, simply rearranged.

For the specific case of a *generalized Fibonacci sequence*, the two **common** ratios are known to be $r = \tau$ and $r' = \tau'$ (quantities defined in Chapter 5), but as mentioned above, these do not fully determine the double geometric sequence which represents a particular generalized Fibonacci sequence, and two more numbers need to be supplied: the values of a_1 and a_2 in the formulas

$$G_i = a_1 \tau^i + a_2 \tau'^i.$$

The derivations above give

$$G_n = (G_1 - G_0 \tau')\tau^n/(\tau - \tau') + (G_0 \tau - G_1)\tau'^n/(\tau - \tau'),$$

and

$$G_n = \tau^{n-1}(G_2 - G_1 \tau')/(\tau - \tau') + \tau'^{n-1}(G_1 \tau - G_2)/(\tau - \tau'),$$

Some alternative forms may be useful, so for the time being, retain the notations

$$a_1 = (G_1 - G_0 \tau')/(\tau - \tau'),$$

$$a_2 = (G_0 \tau - G_1)/(\tau - \tau').$$

Since $\tau' = 1 - \tau$, the expression $\tau - \tau'$ is identical to $2\tau - 1$, so these can also be written as

$$a_1 = (G_1 - G_0 \tau')/(2\tau - 1),$$

$$a_2 = (G_0 \tau - G_1)/(2\tau - 1).$$

For the second case:

$$a_1 = (G_2 - G_1 \tau')/(\tau^2 + 1)$$

$$= (G_2 - G_1 \tau')/(\tau + 2),$$

$$a_2 = (G_2 - G_1 \tau)/(\tau'^2 + 1)$$

$$= (G_2 - G_1 \tau)/(\tau' + 2).$$

Note that in simplifying these equations, the facts that $\tau \tau' = -1$ and $\tau^2 = \tau + 1$ have been used.

It is useful to consolidate all these expressions, for further reference:

For a recurrent sequence A:

$$a_1 = (A_1 - A_0 r')/(r - r')$$
$$= (A_2 - A_1 r')/(r - r')r$$
$$= (k_2 A_0 + r A_1)/(r - r')r$$
$$a_2 = (A_0 r - A_1)/(r - r')$$
$$= (A_1 r - A_2)/(r - r')r'$$
$$= (k_2 A_0 - r' A_1)/(r' - r)r'$$

where

$$r = [k_1 + \sqrt{(k_1^2 + 4k_2)}]/2 \text{ and}$$
$$r' = [k_1 - \sqrt{(k_1^2 + 4k_2)}]/2.$$

For a generalized Fibonacci sequence G:

$$a_1 = (G_1 - G_0 \tau')/(\tau - \tau')$$
$$= (G_1 - G_0 \tau')/(2\tau - 1)$$
$$= (G_0 + G_1 \tau)/(\tau + 2)$$
$$= (G_2 - G_1 \tau')/(\tau + 2),$$
$$a_2 = (G_0 \tau - G_1)/(\tau - \tau')$$
$$= (G_0 \tau - G_1)/(2\tau - 1)$$
$$= (G_0 + G_1 \tau')/(\tau' + 2)$$
$$= (G_2 - G_1 \tau)/(\tau' + 2).$$

It should be noted that there are alternative expressions for a_1 and a_2 above, which can be obtained by expressing τ and τ' in terms of their definitions:

$$2\tau - 1 = 2(1 + \sqrt{5})/2 - 1$$
$$= (1 + \sqrt{5}) - 1$$
$$= \sqrt{5}, \text{ so } 1/(2\tau - 1) = \sqrt{5}/5.$$

Thus

$$a_1 = (G_1 - G_0 \tau')\sqrt{5}/5$$
$$= [G_1 - G_0(1 - \sqrt{5})/2]\sqrt{5}/5$$
$$= [G_1 - (G_0 - G_0\sqrt{5})/2]\sqrt{5}/5$$
$$= (G_1 - G_0/2 + G_0\sqrt{5}/2)\sqrt{5}/5$$

$$= G_0/2 + (G_1 - G_0/2)\sqrt{5}/5$$

and

$$a_2 = (G_0\tau - G_1)\sqrt{5}/5$$
$$= [G_0(1 + \sqrt{5})/2 - G_1]\sqrt{5}/5$$
$$= [(G_0 + G_0\sqrt{5})/2 - G_1]\sqrt{5}/5$$
$$= (G_0/2 + G_0\sqrt{5}/2 - G_1)\sqrt{5}/5$$
$$= G_0/2 + (G_0/2 - G_1)\sqrt{5}/5.$$

The interesting thing about these expressions is that they are identical except for the sign of the coefficient of $\sqrt{5}$ (which, of course, is true of τ and τ' as well).

Any of these expressions can be used for a_1 and a_2 in the equation

$$A_i = a_1 r^i + a_2 r'^i$$

or

$$G_i = a_1\tau^i + a_2\tau'^i,$$

to constitute an **explicit** formula for any recurrent sequence or generalized Fibonacci sequence that has been expressed by a recursive definition.

The expressions for generalized Fibonacci sequences given above can, of course, be applied to the specific cases of the **Fibonacci** and **Lucas** sequences. For the Fibonacci sequence:

$$F_i = a_1\tau^i + a_2\tau'^i$$

with

$$a_1 = (F_0 + F_1\tau)/(\tau + 2)$$

and

$$a_2 = (F_0\tau - F_1)/(2\tau - 1).$$

But $F_0 = 0$ and $F_1 = 1$, so these reduce to

$$a_1 = \tau/(\tau + 2)$$

and

$$a_2 = -1/(2\tau - 1).$$

But a few paragraphs earlier, it was shown that $(2\tau - 1)\tau = \tau + 2$, so the expression for a_1 can be re-written as

$$a_1 = 1/(2\tau - 1)$$

and since τ was defined as $(1 + \sqrt{5})/2$, the expression $2\tau - 1$ can be replaced by $\sqrt{5}$, so

$$F_i = (\tau^i - \tau'^i)/\sqrt{5}, \text{ or}$$

$$F_i = [(\tau^i - \tau'^i)\sqrt{5}]/5.$$

This formula for the numbers in the Fibonacci sequence is sometimes called the **Binet formula,** in honor of **Jacques** Phillippe Marie Binet, the nineteenth-century French mathematician who first derived it. In other books, this formula may be derived differently, and even written in slightly different ways, though the forms found there can be shown to be equivalent to this one. (It might seem strange that a sequence of all integers is obtainable from a formula with the irrational number $\sqrt{5}$ in it, but strangely, when you perform the calculations, all the $\sqrt{5}$'s drop out! The reason is that when you take τ to any power, you have a number of terms which are purely integers and others that are an integer times $\sqrt{5}$; when you take τ' to the same power, you get exactly the same expression with only the signs of the $\sqrt{5}$ terms reversed, so subtracting one from the other leaves only the terms in $\sqrt{5}$, and combining them and dividing by $\sqrt{5}$ gives a pure integer.) It might be useful to use the term **Binet-like formula** for all the general formulas of this chapter in which recurrent sequences and generalized Fibonacci sequences are expressed as double geometric sequences. (Though, as was stated earlier, the term "double geometric sequence" is not found elsewhere, the Binet formula for the Fibonacci sequence and a Binet-like formula for the Lucas sequence can be found elsewhere, but obviously without the context that would show that they are special cases of the representation of recurrent sequences as multiple geometric sequences.) And as was shown above, the a_1 and a_2 coefficients in the formulas for the Binet-like formula expansions for generalized Fibonacci sequences have the signs of $\sqrt{5}$ reversed and are otherwise alike, which explains why, for them as well, all the $\sqrt{5}$ terms drop out. Note that for large i, as was stated in Chapter 4, the τ^i term dominates the expression, so that when i is sufficiently large, the value of F_i is very close to $\tau^i\sqrt{5}/5$. (How large is "sufficiently large"? This depends on how closely you want to approximate F_i. For example, $\tau^4\sqrt{5}/5 = 3.06$, already pretty close to $F_4 = 3$.) This is the explanation for the well-known fact that F_{i+1}/F_i is approximately equal to τ for large i, getting closer as i becomes larger.

For the Lucas sequence, the same procedure can be followed, starting with

$$L_i = a_1\tau^i + a_2\tau'^i$$

with

$$a_1 = (L_0 + L_1\tau)/(\tau + 2)$$

and

$$a_2 = (L_0\tau - L_1)/(2\tau - 1).$$

But $L_0 = 2$ and $L_1 = 1$, so these reduce to

$$a_1 = (2 + \tau)/(\tau + 2) = 1$$

and

$$a_2 = (2\tau - 1)/(2\tau - 1) = 1.$$

So in this case, the formula gives the simpler-looking

$$L_i = \tau^i + \tau^{-i}.$$

In both cases, the powers of τ are large enough that the term in τ^{-i} can be omitted and a close enough approximation to F_i and L_i is obtainable by using the τ^i term alone. Table 2 below illustrates this. For even fairly small values of i, F_i is approximated quite closely by $\tau^i/\sqrt5$, and L_i is approximated quite closely by τ^i.

i	F_i	$\tau^i/\sqrt5$	L_i	τ^i
0	0	0.4472136	2	1
1	1	0.7236068	1	1.6180340
2	1	1.1708204	3	2.6180340
3	2	1.8944272	4	4.2360680
4	3	3.0652476	7	6.8541020
5	5	4.9596748	11	11.0901699
6	8	8.0249224	18	17.9442719
7	13	12.9845971	29	29.0344419
8	21	21.0095195	47	46.9787138
9	34	33.9941166	76	76.0131556
10	55	55.0036361	123	122.9918694
11	89	88.9977528	199	199.0050250
12	144	144.0013889	322	321.9968944
13	233	232.9991416	521	521.0019194
14	377	377.0005305	843	842.9988138
15	610	609.9996721	1364	1364.0007331
16	987	987.0002026	2207	2206.9995469
17	1597	1596.9998748	3571	3571.0002800
18	2584	2584.0000774	5778	5777.9998269
19	4181	4180.9999522	9349	9349.0001070
20	6765	6765.0000296	15127	15126.9999339
21	10946	10945.9999817	24476	24476.0000409

i	F_i	$\tau^i/\sqrt{5}$	L_i	τ^i
22	17711	17711.0000113	39603	39602.9999747
23	28657	28656.9999930	64079	64079.0000156
24	46368	46368.0000043	103682	103681.9999904
25	75025	75024.9999973	167761	167761.0000060
26	121393	121393.0000016	271443	271442.9999963
27	196418	196417.9999990	439204	439204.0000023
28	317811	317811.0000006	710647	710646.9999986

Table 2: Comparison of powers of τ and members of the Fibonacci and Lucas sequences.

Interestingly, since $F_i \approx \tau^i/\sqrt{5}$, and $L_i \approx \tau^i$, it would appear that $F_i L_i \approx F_{2i}$. In fact, the equality is *exact*, not merely *approximate*; one can check the first few values in Table 2 and see this! (And if you wish to see something which is more of a *proof*, just multiply the **Binet** expansion of F_i by the **Binet**-like expansion of L_i.) Another way of looking at how closely F_i is approximated by $\tau^i/\sqrt{5}$, and L_i by τ^i, is to look at the base-τ logarithms of F_i and L_i. For those who do not understand logarithms, the base-τ logarithm of F_i is simply the value of x such that $\tau^x = F_i$. Similarly, the base-τ logarithm of L_i is simply the value of x such that $\tau^x = L_i$. It will be seen that the base-τ logarithms of L_i are very close to integers (in fact, specifically very close to i) for all but the smallest values of i, while the base-τ logarithms of F_i are less than an integer by almost a constant for all but the smallest values of i (specifically, they are very nearly all $i - 1.6722459$). Table 3 shows this.

i	F_i	Base-τ logarithm of F_i	L_i	Base-τ logarithm of L_i
0	0		2	1.44
1	1	0.0000000	1	0.0000000
2	1	0.0000000	3	2.2830118
3	2	1.4404201	4	2.8808402
4	3	2.2830118	7	4.0437704
5	5	3.3445519	11	4.9830348
6	8	4.3212603	18	6.0064437
7	13	5.3301877	29	6.9975334
8	21	6.3267823	47	8.0009414
9	34	7.3280837	76	8.9996403
10	55	8.3275867	123	10.0001374
11	89	9.3277765	199	10.9999475
12	144	10.3277040	322	12.0000200
13	233	11.3277317	521	12.9999923
14	377	12.3277211	843	14.0000029
15	610	13.3277252	1364	14.9999989

i	F_i	Base-τ logarithm of F_i	L_i	Base-τ logarithm of L_i
16	987	14.3277236	2207	16.0000004
17	1597	15.3277242	3571	16.9999998
18	2584	16.3277240	5778	18.0000001
19	4181	17.3277241	9349	19.0000000
20	6765	18.3277241	15127	20.0000000
21	10946	19.3277241	24476	21.0000000
22	17711	20.3277241	39603	22.0000000
23	28657	21.3277241	64079	23.0000000
24	46368	22.3277241	103682	24.0000000
25	75025	23.3277241	167761	25.0000000
26	121393	24.3277241	271443	26.0000000
27	196418	25.3277241	439204	27.0000000
28	317811	26.3277241	710647	28.0000000

Table 3: Base-τ logarithms of members of the Fibonacci and Lucas sequences.

Chapter 11 will look into yet another property of the Fibonacci sequence that illustrates how closely it resembles a geometric progression with common ratio τ, by showing that the difference $F_n - \tau F_{n-1}$ becomes small as n increases.

Chapter 7:   Term-by-term products of recurrent sequences.

While elsewhere in this book, the term "recurrent sequence" is generally understood to refer to 3-term recurrent sequences only, in this and the following chapter (and in Chapter 14, which depends on this chapter), recurrent sequences with higher numbers of terms in the recursion formula will be treated as well. Because every recurrent sequence is also a multiple geometric sequence, it is relatively easy to show that if two recurrent sequences A and A' are multiplied term by term to form a new sequence A'' (which means that $A_i'' = A_i A_i'$) the resulting product sequence is recurrent (but not necessarily with the same number of terms). For the specific case of A and A' being *3-term* recurrent sequences, just write each as a double geometric sequence:

$$A_i = a_1 r^i + a_2 r'^i$$

$$A'_i = a'_1 r''^i + a'_2 r'''^i,$$

and multiply them together to give

$$A_i'' = A_i A_i' = (a_1 r^i + a_2 r'^i)(a'_1 r''^i + a'_2 r'''^i)$$

$$= a_1 a'_1 r^i r''^i + a_1 a'_2 r^i r'''^i + a_2 a'_1 r'^i r''^i + a_2 a'_2 r'^i r'''^i$$

$$= a_1 a'_1 (rr'')^i + a_1 a'_2 (rr''')^i + a_2 a'_1 (r'r'')^i + a_2 a'_2 (r'r''')^i,$$

which is clearly a quadruple geometric sequence, and therefore can be written as a 5-*term* (or fewer) recurrent sequence.

As it is more usual to characterize a n-term recurrent sequence by its first $n - 1$ terms and the recursion coefficients $k_1, ..., k_{n-1}$, this procedure would in most cases be rather unwieldy in practice; first one needs to use the recurrent sequence to double-geometric sequence formulas to get the a's and the r's, then multiply the a's as above to get the new product coefficients, also multiplying the r's to get the common ratios of the four geometric progressions, and then finally use the equivalent to the double-geometric sequence to recurrent sequence formulas (but for quadruple geometric sequences, which makes for a much more complex set of formulas!) to produce the resulting five-term recurrent sequence. It can certainly be done, but it is certainly not an easy task.

When two recurrent sequences are *co-recurrent*, however, the task is much easier. First of all, since the r's depend only on the k's in the above formulas, it is clear that $r = r''$ and $r' = r'''$, so the two middle terms of the four can be combined, meaning that the product sequence is only a 4-term recurrent sequence, and it is relatively easy to compute the necessary k's from the original parameters of A and A'

(without converting them to double geometric sequence coefficients and ratios). This will now be done.

It is known that $A_3 = k_1A_2 + k_2A_1$, $A'_3 = k_1A'_2 + k_2A'_1$, $A_4 = k_1A_3 + k_2A_2$, and $A'_4 = k_1A'_3 + k_2A'_2$, with the same k_1 and k_2 for both sequences, since they were specified as co-recurrent. Since $A_i'' = A_iA_i'$ for all i,

$$A_1'' = A_1A_1',$$

$$A_2'' = A_2A_2',$$

$$A_3'' = A_3A_3'$$

$$= (k_1A_2 + k_2A_1)(k_1A'_2 + k_2A'_1)$$

$$= k_1^2A_2A'_2 + k_1k_2(A_1A'_2 + A_2A'_1) + k_2^2A_1A'_1, \text{ and}$$

$$A_4'' = A_4A_4'$$

$$= (k_1A_3 + k_2A_2)(k_1A'_3 + k_2A'_2)$$

$$= [k_1(k_1A_2 + k_2A_1) + k_2A_2][k_1(k_1A'_2 + k_2A'_1) + k_2A'_2]$$

$$= (k_1^2A_2 + k_1k_2A_1 + k_2A_2)(k_1^2A'_2 + k_1k_2A'_1 + k_2A'_2)$$

$$= [(k_1^2 + k_2)A_2 + k_1k_2A_1][(k_1^2 + k_2)A'_2 + k_1k_2A'_1]$$

$$= (k_1^2 + k_2)^2A_2A'_2 + k_1k_2(k_1^2 + k_2)(A_1A'_2 + A_2A'_1) + k_1^2k_2^2A_1A'_1.$$

To establish a 4-term recurrence relation, it is necessary to find three coefficients k'_1, k'_2, and k'_3 such that

$$A_4'' = k'_1A_3'' + k'_2A_2'' + k'_3A_1''.$$

To satisfy this equation, three conditions are sufficient:

$$(k_1^2 + k_2)^2 = k'_1k_1^2 + k'_2,$$

$$k_1k_2(k_1^2 + k_2) = k'_1k_1k_2, \text{ and}$$

$$k_1^2k_2^2 = k'_1k_2^2 + k'_3.$$

From the second of these it is immediately obvious that

$$k'_1 = k_1^2 + k_2,$$

and by substituting this value into each of the others, it is easy to obtain

$$k'_2 = (k_1^2 + k_2)^2 - k'_1k_1^2$$

$$= (k_1^2 + k_2)^2 - k_1^2(k_1^2 + k_2)$$

$$= (k_1^4 + 2k_1^2k_2 + k_2^2) - (k_1^4 + k_1^2k_2)$$

$$= k_1^2k_2 + k_2^2$$

$$= (k_1^2 + k_2)k_2, \text{ and}$$

$$k'_3 = k_1^2 k_2^2 - k'_1 k_2^2$$

$$= k_1^2 k_2^2 - k_2^2(k_1^2 + k_2)$$

$$= k_1^2 k_2^2 - (k_1^2 k_2^2 + k_2^3)$$

$$= -k_2^3.$$

Since all the equations used in this derivation will be identically satisfied if A_1, A_2, A_3, A_4, A'_1, A'_2, A'_3, and A'_4 are replaced by any four consecutive terms of the two sequences, the recurrence relation will hold for all the rest of the terms of A'' with those values of k'_1, k'_2, and k'_3. Also, since none of the A or A' terms appear in any of the expressions for the k' terms, either A or A' may be replaced by any other sequence co-recurrent with it, and in fact, even replacing both A and A' by sequences co-recurrent with them would leave the k' terms unchanged. While this has here only been demonstrated for both A and A' being 3-term recurrent sequences, it is a general rule, which however will not be demonstrated here because to do so would involve more mathematical manipulations than the reader is likely to want to follow. It will just be stated that:

- **If two (or more) recurrent sequences are multiplied term by term to produce a new recurrent sequence, and then any of the original sequences is replaced by a different one co-recurrent with it, the resulting product sequence is co-recurrent with the original product sequence.**

It might also be noted that nothing in the above argument requires A and A′ to be distinct sequences. A sequence is certainly co-recurrent with itself according to the definition of co-recurrence given in this book, so one can see that the squares of any 3-term recurrent sequence A, whose recursion coefficients are k_1 and k_2, form a 4-term recurrent sequence whose recursion coefficients are $k_1^2 + k_2$, $(k_1^2 + k_2)k_2$, and $-k_2^3$. And so do products of terms of a recurrent sequence separated by the same distance; e. g., A_1A_5, A_2A_6, A_3A_7, etc.

Furthermore, by putting 1 for both k_1 and k_2, it can be seen that the squares of the terms of the Fibonacci sequence, the Lucas sequence, or any generalized Fibonacci sequence form a 4-term recurrent sequence whose recursion coefficients are 2, 2, and –1, as do the term-by-term products of any two of these. Table 4 shows this for the squares of terms of the Fibonacci sequence, but it can be done for the other cases as well.

i	F_i	F_i^2	F_{i-1}^2	F_{i-2}^2	F_{i-3}^2	2a + 2b – c	
			=a	=b	=c		
0	0	0	1	1	4	2(1)+2(1)-(4)	0
1	1	1	0	1	1	2(0)+2(1)-(1)	1

i	F_i	F_i^2	F_{i-1}^2	F_{i-2}^2	F_{i-3}^2	$2a + 2b - c$	
			=a	=b	=c		
2	1	1	1	0	1	2(1)+2(0)-(1)	1
3	2	4	1	1	0	2(1)+2(1)-(0)	4
4	3	9	4	1	1	2(4)+2(1)-(1)	9
5	5	25	9	4	1	2(9)+2(4)-(1)	25
6	8	64	25	9	4	2(25)+2(9)-(4)	64
7	13	169	64	25	9		169
8	21	441	169	64	25		441
9	34	1156	441	169	64		1156
10	55	3025	1156	441	169		3025
11	89	7921	3025	1156	441		7921
12	144	20736	7921	3025	1156	etc.	20736
13	233	54289	20736	7921	3025		54289
14	377	142129	54289	20736	7921		142129
15	610	372100	142129	54289	20736		372100
16	987	974169	372100	142129	54289		974169
17	1597	2550409	974169	372100	142129		2550409
18	2584	6677056	2550409	974169	372100		6677056

Table 4: Four-term recurrence relation of squares of Fibonacci sequence terms.

The fact that the term-by-term product of two generalized Fibonacci sequences (or the term-by-term square of any generalized Fibonacci sequence) is a 4-term recurrent sequence whose recursion coefficients are 2, 2, and –1, perhaps, should warrant a new name for such recurrent sequences. None is known to me, so a name will be invented in this book: a ***generalized Fibonacci sequence pairwise product sequence.*** A generalized Fibonacci sequence pairwise product sequence can be considered to be defined either as a term-by-term product of two generalized Fibonacci sequences (including, as a special case, the term-by-term square of any generalized Fibonacci sequence) or as a 4-term recurrent sequence whose recursion coefficients are 2, 2, and –1. One can similarly speak of generalized Fibonacci sequence triplewise product sequences, of generalized Fibonacci sequence quadruplewise product sequences, of generalized Fibonacci sequence quintuplewise product sequences, etc. If one wishes to talk of the general case, the term "generalized Fibonacci sequence *n*-tuplewise product sequences" can be used.

Although this chapter is mainly concerned with term-by-term products of *two* recurrent sequences, it might be noted that it is possible to consider larger numbers of sequences multiplied together on a term-by-term basis. And, just as using generalized Fibonacci sequences as the two recurrent sequences gives a regular set of recursion coefficients (2, 2, and –1), if *three* generalized Fibonacci sequences are multiplied term by term (including the special case of *cubing* a generalized Fibonacci sequence term by term) there will always result a recurrent sequence with the same recursion coefficients (3, 6, –3, and –1), for example

$$3(2^3) + 6(1^3) - 3(1^3) - 1(0^3) = 3(8) + 6(1) - 3(1) - 1(0) = 27 = 3^3,$$

$$3(3^3) + 6(2^3) - 3(1^3) - 1(1^3) = 3(27) + 6(8) - 3(1) - 1(1) = 125 = 5^3,$$

$$3(5^3) + 6(3^3) - 3(2^3) - 1(1^3) = 3(125) + 6(27) - 3(8) - 1(1) = 512 = 8^3,$$

etc.

Similarly,

$$G_n{}^4 = 5G_{n-1}{}^4 + 15G_{n-2}{}^4 - 15G_{n-3}{}^4 - 5G_{n-4}{}^4 + G_{n-5}{}^4,$$

$$G_n{}^5 = 8G_{n-1}{}^5 + 40G_{n-2}{}^5 - 60G_{n-3}{}^5 - 40G_{n-4}{}^5 + 8G_{n-5}{}^5 + G_{n-6}{}^5,$$

and so forth. It would not be difficult to form the conclusion that, for any set of generalized Fibonacci sequences, multiplying them together term by term would always produce a set of recursion coefficients that depends only on the number of sequences taken. (And this, of course, also extends to taking the powers of the terms of a single generalized Fibonacci sequence as the terms of a new recurrent sequence.)

This line will be pursued further, but because some additional concepts are useful that are yet to be introduced in Chapter 13, the discussion will resume after that chapter.

Chapter 8:   Analyzing generalized Fibonacci sequence pairwise product sequences.

In Chapter 7, the term *generalized Fibonacci sequence pairwise product sequence* was introduced to refer to 4-term recurrent sequences which have the specific recursion coefficients 2, 2, and –1. The question arises, given such a sequence, is it always possible to find two generalized Fibonacci sequences whose term-by-term product is the specified sequence? And so, consider attempting to do exactly this. Let the given recurrent sequence be represented by A, where it is known that

$$A_3 = 2A_2 + 2A_1 - A_0,$$

$$A_4 = 2A_3 + 2A_2 - A_1,$$

$$A_5 = 2A_4 + 2A_3 - A_2,$$

$$A_6 = 2A_5 + 2A_4 - A_3, \text{ etc.}$$

The goal is to find two generalized Fibonacci sequences G and G' such that $A_i = G_i G'_i$ for all i. Clearly it would be required that $A_0 = G_0 G'_0$ and $A_1 = G_1 G'_1$. Two additional conditions would need to be supplied to fix G_0, G'_0, G_1, and G'_1. One of these is set by requiring $A_2 = G_2 G'_2$, since the specification of G and G' as generalized Fibonacci sequences implies that $G_2 = G_0 + G_1$ and $G'_2 = G'_0 + G'_1$. This means that

$$A_2 = G_2 G'_2$$

$$= (G_0 + G_1)(G'_0 + G'_1)$$

$$= G_0 G'_0 + G_0 G'_1 + G_1 G'_0 + G_1 G'_1,$$

which, combined with the facts that $A_0 = G_0 G'_0$ and $A_1 = G_1 G'_1$, imply that

$$A_2 = A_0 + G_0 G'_1 + G_1 G'_0 + A_1, \text{ or}$$

$$A_2 - A_1 - A_0 = G_0 G'_1 + G_1 G'_0.$$

This gives only three equations to solve for the four quantities G_0, G'_0, G_1, and G'_1. And attempting to use $A_3 = G_3 G'_3$ leads to nothing new. For putting

$$A_3 = G_3 G'_3$$

$$= (G_1 + G_2)(G'_1 + G'_2)$$

$$= [G_1 + (G_0 + G_1)][G'_1 + (G'_0 + G'_1)]$$

$$= (G_0 + 2G_1)(G'_0 + 2G'_1)$$

$$= G_0 G'_0 + 2G_0 G'_1 + 2G_1 G'_0 + 4G_1 G'_1$$

$$= A_0 + 2(A_2 - A_1 - A_0) + 4A_1$$

$$= A_0 + 2A_2 - 2A_1 - 2A_0 + 4A_1$$

$$= 2A_2 + 2A_1 - A_0,$$

which simply restates the recurrence relation for A, so it sets no further condition. So does this mean that any G_0, G'_0, G_1, and G'_1 such that $G_0G'_0 = A_0$, $G_1G'_1 = A_1$, and $G_0G'_1 + G_1G'_0 = A_2 - A_1 - A_0$ will do?

In fact, the answer is *yes*. Suppose G_0 to be fixed, for example. From this can be calculated $G'_0 = A_0/G_0$. Whatever G_1 is, $G'_1 = A_1/G_1$. So the equation $G_0G'_1 + G_1G'_0 = A_2 - A_1 - A_0$ becomes

$$G_0A_1/G_1 + G_1A_0/G_0 = A_2 - A_1 - A_0, \text{ or}$$

$$G_0^2A_1 + G_1^2A_0 = G_0 G_1(A_2 - A_1 - A_0).$$

With G_0, A_0, A_1, and A_2 all determined, this is a quadratic equation in G_1, and it turns out that either value of the solution can be used. Consider a sample sequence A, with $A_0 = 12$, $A_1 = 6$, and $A_2 = 35$. By applying the recursion formula $A_{n+3} = 2A_{n+2} + 2A_{n+1} - A_n$, the next few terms can be calculated as $A_4 = 70$, $A_5 = 204$, etc. If one writes x for G_1/G_0, the equation

$$G_0^2A_1 + G_1^2A_0 = G_0G_1(A_2 - A_1 - A_0)$$

can be written as

$$A_1 + A_0x^2 = (A_2 - A_1 - A_0)x,$$

which rearranges to

$$A_0x^2 + (A_0 + A_1 - A_2)x + A_1 = 0,$$

or in this case

$$12x^2 - 17x + 6 = 0,$$

whose solutions are $x = 3/4$ and $x = 2/3$. This really means that

$$G_1 = 3G_0/4, \ G'_0 = 12/G_0, \ G'_1 = 8/G_0, \text{ or}$$

$$G_1 = 2G_0/3, \ G'_0 = 12/G_0, \ G'_1 = 9/G_0.$$

Note that this means that all the terms of G can be multiplied by a constant, with the terms of G' divided by that same constant. Furthermore, the two choices of x really imply an interchange of the two sequences. This can be seen by looking at the cases $x = 3/4$ with $G_0 = 12$ and $x = 2/3$ with $G_0 = 1$. For the first case,

$$G_1 = 3(12)/4 = 9,$$

$$G'_0 = 12/12 = 1, \text{ and}$$

$$G'_1 = 8/12 = 2/3,$$

while for the second,

$$G_1 = 2/3,$$

$$G'_0 = 12/1 = 12, \text{ and}$$

$$G'_1 = 9/1 = 9.$$

These are exactly the same numbers, except that the two sequences are interchanged. So in fact, the fact that only three conditions are available to calculate the four required variables G_0, G'_0, G_1, and G'_1 is not, actually, a problem. This freedom is real.

Chapter 9:   Co-recurrent sequences as a vector space.

It is obvious that multiplying the terms of a recurrent sequence by a constant gives a recurrent sequence with the *same recursion formula*. For example, if the recurrent sequence A is a 3-term recurrent sequence such that $A_i = k_1 A_{i-1} + k_2 A_{i-2}$, and a second sequence A' is defined with $A'_i = cA_i$, then

$$A'_i = cA_i$$

$$= c(k_1 A_{i-1} + k_2 A_{i-2})$$

$$= ck_1 A_{i-1} + ck_2 A_{i-2}$$

$$= k_1 A'_{i-1} + k_2 A'_{i-2}.$$

(Exactly the same argument holds for recurrent sequences with different numbers of terms in the recursion formula, but as this book is mainly concerned with 3-term recurrent sequences, this type was used in the example.)

It is also easy to see that adding two recurrent sequences with the same recursion formula (*i. e., co-recurrent sequences*, as defined in Chapter 3) gives yet another recurrent sequence with the *same* **recursion** formula. For example, if two 3-term recurrent sequences A and A' are defined such that $A_i = k_1 A_{i-1} + k_2 A_{i-2}$ and $A'_i = k_1 A'_{i-1} + k_2 A'_{i-2}$, and a third sequence A'' is defined such that $A''_i = A'_i + A_i$, then

$$A''_i = A'_i + A_i$$

$$= (k_1 A_{i-1} + k_2 A_{i-2}) + (k_1 A'_{i-1} + k_2 A'_{i-2})$$

$$= (k_1 A_{i-1} + k_1 A'_{i-1}) + (k_2 A_{i-2} + k_2 A'_{i-2})$$

$$= k_1 (A_{i-1} + A'_{i-1}) + k_2 (A_{i-2} + A'_{i-2})$$

$$= k_1 (A''_{i-1}) + k_2 (A''_{i-2}).$$

Because recurrent sequences have this property, one can write $A' = cA$ to mean $A'_i = cA_i$ for all i, and $A'' = A' + A$ to mean $A''_i = A'_i + A_i$ for all i. When written in this way, so one can talk of multiplying a whole sequence by a constant or adding two sequences together, the two properties just demonstrated imply that recurrent sequences with the same recursion formula form what is known to mathematicians as a *vector space*. (Actually, a number of other properties are required to constitute a vector space, and it can be shown that co-recurrent sequences satisfy these requirements, but they are not important in this discussion.) Vector spaces are so important to mathematicians that a whole branch of mathematics, known as *linear algebra*, has developed to deal with them. Linear algebra turns out to be an important part of both pure mathematics (mathematics for its own sake) and applied mathematics (mathematics used in the sciences, especially in this case physics and chemistry). While this book will not discuss linear algebra at any length, the name "standard co-recurrent basis sequences" was chosen because in the vector space just described, they form a **basis**, in the sense in which the term is used in linear algebra, for the space. (As anyone who has studied linear algebra would know, of course, there are an infinite number of ways of choosing a basis for a vector space. It is for this reason that, at least within the confines of this book, *this* basis is specially singled out by the term "standard.") Rather than trying to presenting a course in linear algebra, however, this book will instead refer the reader to any text on the subject if he or she wants to follow up on this subject.

It will be noted that the second of these two properties was stated as "adding two recurrent sequences with the *same recursion formula* gives yet another recurrent sequence with the same **recursion** formula." The reader may ask what happens when two recurrent sequences with *different* recursion formulas are added. The result is always a recurrent sequence, but whose recursion formula is not related to the recursion formulas of the individual sequences; it may even have a *different number of terms* from either of the recursion formulas of the individual sequences. (This has already been seen. A geometric progression is a sequence that can be written as a *2-term* recurrent sequence, but adding two geometric progressions with different common ratios gives a double geometric sequence, which can be written as a *3-term* recurrent sequence.) Most of the recurrent sequences discussed in this book are generalized Fibonacci sequences, of course, and *all* generalized Fibonacci sequences, by definition, share a common recursion formula. Most of the time, even when discussing different recurrent sequences which are not generalized Fibonacci sequences together, they will be co-recurrent.

Chapter 10: Limiting ratios of recurrent sequence terms.

Consider two different recurrent sequences A and A'. Each can be written, as shown in Chapter 6, as a double geometric sequence. So it is possible to write

$$A_i = a_1 r^i + a_2 r'^i,$$

$$A'_i = a_1' r''^i + a_2' r'''^i.$$

The decision as to which of the two geometric progression common ratios will be designated as r and which as r' will be specifically on the basis of largest absolute value: they will be chosen so that $|r| > |r'|$; similarly, the choice of which of the two geometric progression common ratios will be designated as r'' and which as r''' will be made so that $|r''| > |r'''|$. The arguments of Chapter 4 can then be used to show that (when i is large enough) one can put $A_i \approx a_1 r^i$ and $A'_i \approx a_1' r''^i$, so

$$A_i/A'_i \approx a_1 r^i / a_1' r''^i = (a_1/a_1')(r/r'')^i.$$

The result of this calculation is:

For two different recurrent sequences A and A', which can be written

$$A_i = a_1 r^i + a_2 r'^i, \ A'_i = a_1' r''^i + a_2' r'''^i$$

where $|r| > |r'|$ and $|r''| > |r'''|$, as i becomes very large, the ratio A_i/A'_i:

1.   approaches the ratio (a_1/a_1') if $r = r''$,

2.   approaches 0 if $r < r''$, and

3.   diverges to infinity if $r > r''$.

Of course, for two co-recurrent sequences, the first of these three options applies. And since any two recurrent sequences related by a shift are co-recurrent, this implies that for any recurrent sequence, as i becomes very large, the ratio A_{i+1}/A_i approaches the ratio r. (For any *generalized Fibonacci sequence*, this limiting ratio is τ.) And if the sequences are specified by recursive definitions, it will be necessary to apply the recurrent sequence to double-geometric sequence formulas of Chapter 4 to use this rule. In particular, when r and r' have been chosen to make $|r| > |r'|$, one only needs to use $a_1 = (A_1 - A_0 r')/(r - r')$, as the terms in a_2 are unused in this discussion. Also, for all generalized Fibonacci sequences, it is known that $r = \tau$ and $r' = \tau'$, so one can simply state that:

For two different generalized Fibonacci sequences G and G', as i

becomes very large, the ratio G_i/G'_i approaches the ratio

$$(G_1 - G_0\tau')/(G'_1 - G'_0\tau')$$

$$= (G_1\tau + G_0)/(G'_1\tau + G'_0).$$

Table 5 shows how quickly the quotients of terms approach the limiting value for a sample case of two generalized Fibonacci sequences. The specific choice of G and G' in Table 5 (namely $G_1 = 5$, $G_2 = 2$, $G'_1 = 13$, and $G'_2 = 3$) is arbitrary; the same result would be observed for any two generalized Fibonacci sequences that one might choose.

G	G'	G_i/G'_i
5	13	0.38461538
2	3	0.66666667
7	16	0.43750000
9	19	0.47368421
16	35	0.45714286
25	54	0.46296296
41	89	0.46067416
66	143	0.46153846
107	232	0.46120690
173	375	0.46133333
280	607	0.46128501
453	982	0.46130346
733	1589	0.46129641
1186	2571	0.46129911
1919	4160	0.46129808
3105	6731	0.46129847
5024	10891	0.46129832
8129	17622	0.46129838
13153	28513	0.46129836
21282	46135	0.46129836
34435	74648	0.46129836

Table 5: Approach of ratio of two generalized Fibonacci sequences to a limit.

Chapter 11: More about the near-geometric behavior of generalized Fibonacci sequences.

This chapter is mainly concerned with generalized Fibonacci sequences, including the Fibonacci sequence itself, and their relation to the golden ratio τ. It was earlier noted that as n increases, the ratio G_n/G_{n-1} becomes closer and closer to τ. So it might be worthwhile to look at the difference $G_n - \tau G_{n-1}$. This can be best handled by determining the Binet-like formula for G_n:

$$G_n = a_1 \tau^n + a_2 \tau'^n,$$

where

$$a_1 = (G_1 - G_0 \tau') \sqrt{5}/5$$

and

$$a_2 = (G_0 \tau - G_1) \sqrt{5}/5,$$

or, with $n - 1$ for n,

$$G_{n-1} = a_1 \tau^{n-1} + a_2 \tau'^{n-1}.$$

Making use of the fact that $\tau\tau' = -1$, this means that

$$\tau G_{n-1} = a_1 \tau^n - a_2 \tau'^{n-2}, \text{ or}$$

$$G_n - \tau G_{n-1} = a_1 \tau^n + a_2 \tau'^n - (a_1 \tau^n - a_2 \tau'^{n-2})$$

$$= a_2 \tau'^n + a_2 \tau'^{n-2}$$

$$= a_2 \tau'^{n-2}(\tau'^2 + 1).$$

Since $\tau'^2 = \tau' + 1$, this can be written as

$$G_n - \tau G_{n-1} = a_2 \tau'^{n-2}(\tau' + 2).$$

But it is also possible to look at

$$G_{n-2} = a_1 \tau^{n-2} + a_2 \tau'^{n-2}, \text{ which means}$$

$$G_{n-2} + \tau G_{n-1} = a_1 \tau^{n-2} + a_1 \tau^n$$

$$= a_1 \tau^{n-2}(1 + \tau^2),$$

and again $\tau^2 = \tau + 1$, so this can be written as

$$G_{n-2} + \tau G_{n-1} = a_1 \tau^{n-2}(2 + \tau).$$

If these are multiplied together, one obtains

$$(G_n - \tau G_{n-1})(G_{n-2} + \tau G_{n-1}) = a_1 a_2 \tau'^{n-2} \tau^{n-2}(\tau' + 2)(2 + \tau)$$

$$= (-1)^{n-2} a_1 a_2 (\tau' + 2)(2 + \tau)$$

$$= (-1)^n a_1 a_2 (2\tau' + \tau\tau' + 4 + 2\tau),$$

But $\tau' + \tau = 1$ and $\tau\tau' = -1$, so $2\tau' + \tau\tau' + 2\tau = 1$, and

$$(G_n - \tau G_{n-1})(G_{n-2} + \tau G_{n-1}) = 5(-1)^n a_1 a_2.$$

Substituting for a_1 and a_2,

$$(G_n - \tau G_{n-1})(G_{n-2} + \tau G_{n-1}) = 5(-1)^n (G_1 - G_0\tau')(G_0\tau - G_1)/5$$

$$= (-1)^n (G_0 G_1 \tau - G_0^2 \tau\tau' - G_1^2 + G_0 G_1 \tau')$$

$$= (-1)^n (G_0 G_1 \tau + G_0^2 - G_1^2 + G_0 G_1 \tau')$$

$$= (-1)^n [G_0^2 - G_1^2 + G_0 G_1 (\tau + \tau')]$$

$$= (-1)^n (G_0^2 - G_1^2 + G_0 G_1),$$

which (except for the alternating sign from the first factor) does not depend on n. It should be noted that if this equation is written in the form

$$G_n - \tau G_{n-1} = (-1)^n (G_0^2 - G_1^2 + G_0 G_1)/(G_{n-2} + \tau G_{n-1}),$$

the denominator of the right-hand side increases geometrically as n does, so it becomes very clear that $G_n - \tau G_{n-1}$ becomes smaller as n increases.

Since $F_0 = 0$ and $F_1 = 1$, when G is F, the equations become

$$(F_n - \tau F_{n-1})(F_{n-2} + \tau F_{n-1}) = (-1)^{n-1}/5,$$

One might note that since $F_0 = 0$, and $F_1 = F_2 = 1$, this equation simply becomes $(1 - \tau)\tau = -1$ for $n = 2$. Since $1 - \tau = \tau'$, this is certainly consistent with the known fact that $\tau\tau' = -1$, but since both the facts that $1 - \tau = \tau'$ and $\tau\tau' = -1$ were used in deriving the equation, this really does not *prove* anything; it just shows that $\tau\tau' = -1$ is included as a special case of the formula.

The reader might ask why, unlike most of the formulas in this book, this was not worked out for recurrent sequences in general, rather than specifically for generalized Fibonacci sequences. The main reason is that if one begins with the Binet-like formula for A_n:

$$A_n = a_1 r^n + a_2 r'^n,$$

or, with $n - 1$ for n,

$$A_{n-1} = a_1 r^{n-1} + a_2 r'^{n-1},$$

and makes use of the fact that $rr' = -k_2$, (as the corresponding generalized Fibonacci sequence formula $\tau\tau' = -1$ was used above) this gives

$$rA_{n-1} = a_1r^n - k_2a_2r'^{n-2}, \text{ or}$$

$$A_n - rA_{n-1} = a_2r'^n + k_2a_2r'^{n-2}$$

$$= a_2r'^{n-2}(r'^2 + k_2).$$

From $k_1 = r + r'$ and $k_2 = -rr'$ can be derived

$$r = -k_2/r',$$

$$k_1 = -k_2/r' + r',$$

$$k_1r' = -k_2 + r'^2, \text{ and thus}$$

$$r'^2 = k_2 + k_1r'.$$

Consequently, the expression for $A_n - rA_{n-1}$ can be written as

$$A_n - rA_{n-1} = a_2r'^{n-2}(2k_2 + k_1r').$$

But it turns out that if one tries to follow the procedure above to look at $A_{n-2} = a_1r^{n-2} + a_2r'^{n-2}$, in order to get the $a_2r'^{n-2}$ terms to cancel, it is necessary to multiply A_{n-2} by k_2, and then, though one can, analogously with the previous argument, write

$$k_2A_{n-2} + rA_{n-1} = k_2a_1r^{n-2} + a_1r^n$$

$$= a_1r^{n-2}(k_2 + r^2),$$

and again (because r and r' can be interchanged in the earlier derivation) $r^2 = k_2 + k_1r$, so this can be written as

$$k_2A_{n-2} + rA_{n-1} = a_1r^{n-2}(2k_2 + k_1r),$$

If these are multiplied together, one obtains

$$(A_n - rA_{n-1})(k_2A_{n-2} + rA_{n-1}) = [a_2r'^{n-2}(2k_2 + k_1r')][a_1r^{n-2}(2k_2 + k_1r)]$$

$$= a_1a_2r^{n-2}r'^{n-2}(2k_2 + k_1r')(2k_2 + k_1r)$$

$$= a_1a_2(-k_2)^{n-2}(2k_2 + k_1r')(2k_2 + k_1r)$$

$$= a_1a_2(-k_2)^{n-2}(4k_2^2 + 2k_1k_2r' + 2k_1k_2r + k_1^2rr')$$

$$= a_1a_2(-k_2)^{n-2}[4k_2^2 + 2k_1k_2(r' + r) + k_1^2rr'].$$

But $r + r' = k_1$ and $rr' = -k_2$, so this can be rewritten as

$$(A_n - rA_{n-1})(k_2A_{n-2} + rA_{n-1}) = a_1a_2(-k_2)^{n-2}(4k_2^2 + 2k_1^2k_2 - k_1^2k_2)$$

$$= a_1a_2(-k_2)^{n-2}(4k_2^2 + k_1^2k_2)$$

$$= a_1a_2(-k_2)^{n-1}(4k_2 + k_1^2).$$

(Note at this point that the expression $4k_2 + k_1^2$, which occurs inside the last set of parentheses, is the same expression appearing in the radical of the expression for r in terms of k_1 and k_2.) It may be noted that the only factor on the right-hand side of the equation that depends on n is the factor $(-k_2)^{n-1}$. But when one goes on to divide by the expression $(k_2 A_{n-2} + r A_{n-1})$, the final formula becomes

$$A_n - r A_{n-1} = a_1 a_2 (-k_2)^{n-1}(4k_2 + k_1^2)/(k_2 A_{n-2} + r A_{n-1}).$$

And while, for generalized Fibonacci sequences, the analogous formula

$$G_n - \tau G_{n-1} = (-1)^n (G_0^2 - G_1^2 + G_0 G_1)/(G_{n-2} + \tau G_{n-1})$$

had a denominator $(G_{n-2} + \tau G_{n-1})$ on the right-hand side which increased geometrically and a numerator that was constant except for sign, the right-hand side of the expression for $A_n - r A_{n-1}$ has the factor $(-k_2)^{n-1}$ in the numerator, which (in the case where $|k_2| > 1$) can *also* increase geometrically with n. Since it would be beyond the scope of this book to examine all the possibilities that depend on the values of $|k_2|$ and $|r|$, It is necessary to confine the discussion to generalized Fibonacci sequences.

Chapter 12:   Review of recurrent/multiple geometric sequence facts.

So far, a number of facts about recurrent sequences (some specific to 3-term recurrent sequences, others more general) and their equivalent multiple geometric sequences have been brought out. While such things as the exact expressions for converting between the two representations of a sequence given at the end of Chapter 4 are not worth memorizing, some of the general points which have been made up to now should be kept in mind. Later parts of this book will assume them as basic. A summary of these is given below.

1. Every arithmetic progression is a 3-term recurrent sequence. All arithmetic progressions have the same recurrence relation.

2. Every **geometric** progression is a 2-term recurrent sequence.

3. Every multiple **geometric** progression is an n-term recurrent sequence, where n is 1 more than the number of geometric progressions summed term-by-term. The coefficients in the recurrence relation can be computed from the **common** ratios of the geometric progressions summed to produce any multiple geometric progression.

4. Conversely, almost every recurrent sequence is a multiple geometric progression, where where the number of geometric progressions summed term-by-term is 1 less than the number of terms in the recurrent sequence. The common ratios of the geometric progressions summed to produce any recurrent sequence can be computed from the coefficients in the recurrence relation. (The few exceptions include arithmetic progressions, which, as shown in Chapter 3, cannot be expressed as double geometric progressions.)

5. A generalized Fibonacci sequence has been defined as a recurrent sequence, whose recurrence relation is the *same* 3-term recurrence relation for all generalized Fibonacci sequences, and thus is also a double geometric sequence.

6. The **common** ratios of the two geometric progressions summed to produce any generalized Fibonacci sequence are always $r = \tau$ and $r' = \tau'$. These numbers have three important properties which should be memorized, as they are constantly used in this book: $\tau + \tau' = 1$, $\tau\tau' = -1$, and both τ and τ' satisfy the equation $x^2 = x + 1$. (Sometimes the first two of them will be used in different forms, such as $\tau = 1 - \tau'$ or $\tau = -1/\tau'$. All of these should be easily recognized.)

Chapter 13:   Binomial coefficients and generalized Fibonacci sequences.

Some of the readers may be familiar with the **binomial theorem,** that specifies the coefficients in the expansion of the expression

$$(x + y)^n.$$

In that expansion, every term is a constant multiplied by $x^p y^{n-p}$, where the exponents of x and of y sum to n. The coefficient of $x^p y^{n-p}$ is given by

$$_nC_p = n!/p!(n - p)!$$

where n! means the **factorial,** defined as

$$n! = n(n - 1)(n - 2)...1$$

(it should be noted that the symbol $_nC_p$ is derived from the statistics of permutations and combinations, where C stands for "combinations"; other books, such as Dunlap's and Vajda's, use a different notation, $\binom{n}{p}$, which I avoid because in my opinion it is less clear, being easily mistaken for a fraction.)

The formula just presented, of course, is an explicit definition of the binomial coefficients $_nC_p$. Since this book is concerned with recurrent sequences and a chapter is here being devoted to binomial coefficients, it should not be surprising to the reader that a recursive definition of the binomial coefficients $_nC_p$ will be presented at this point. In fact, they can be recursively defined by formulas that closely resemble the definition of the Fibonacci sequence:

$$_nC_0 = 1,$$

$$_nC_n = 1,$$

$$_nC_p = {_{n-1}C_{p-1}} + {_{n-1}C_p}.$$

One other interesting relationship applies, which will not be proved here, but which will be used in this chapter:

$$_nC_p = {_nC_{n-p}}.$$

With just these four formulas, all the binomial coefficients can be calculated, without the need to calculate all the factorials in the earlier formula. For

$$_1C_0 = 1,$$

$$_1C_1 = 1,$$

$$_2C_0 = 1,$$

$$_2C_1 = {}_1C_0 + {}_1C_1 = 1 + 1 = 2,$$

$$_2C_2 = 1,$$

and then all the $_3C_p$ values can be calculated from the $_2C_p$ values, and so forth. The binomial coefficients can be presented in a triangular format, and at least two formats are common. Table 6 shows them in a format which is more convenient for picking out the binomial coefficient $_nC_p$ corresponding to particular values of n and p, while Table 7 illustrates more clearly the symmetrical properties of the binomial coefficients.

p	0	1	2	3	4	5	6	7	8	9
n										
0	1									
1	1	1								
2	1	2	1							
3	1	3	3	1						
4	1	4	6	4	1					
5	1	5	10	10	5	1				
6	1	6	15	20	15	6	1			
7	1	7	21	35	35	21	7	1		
8	1	8	28	56	70	56	28	8	1	
9	1	9	36	84	126	126	84	36	9	1

Table 6: Pascal's triangle of binomial coefficients (with n and p values).

								1										
							1		1									
						1		2		1								
					1		3		3		1							
				1		4		6		4		1						
			1		5		10		10		5		1					
		1		6		15		20		15		6		1				
	1		7		21		35		35		21		7		1			
1		8		28		56		70		56		28		8		1		
1	9		36		84		126		126		84		36		9		1	

Table 7: Pascal's triangle of binomial coefficients (symmetrical format).

These triangular arrangements, especially that of Table 7, are generally known as **Pascal's triangle,** in honor of the French mathematician **Blaise** Pascal, who was the first European to describe it; however, earlier diagrams illustrating this arrangement were found in writings in places like China long before Pascal. Notice that in the arrangement in Table 7, the two numbers immediately above each individual entry (one above and to the left, one above and to the right) add to that number, and the central column is a mirror, with entries on either side the same distance from the center being identical. The formulas of this chapter relate these coefficients $_nC_p$ to recurrent sequences, including, of course, as special cases, generalized Fibonacci sequences.

It might be noted that the column headed by $p = 0$ is all 1's, the column headed by $p = 1$ is simply all the integers in order, and the column headed by $p = 2$ contains the so-called *triangular numbers*, namely the number of objects that can be arranged in a triangular array. It is less obvious, but the diagonal arrays of cells with $n - p = 0$, 1, or 2 give the same numerical values (because of the symmetry of the rows).

Noting that $_1C_0$ and $_1C_1$ are both 1, the standard **recursion** formula defining a recurrent sequence,

$$A_n = k_1 A_{n-1} + k_2 A_{n-2},$$

or alternatively,

$$A_{n+2} = k_1 A_{n+1} + k_2 A_n,$$

can be put into the form

$$A_{n+2} = k_1(_1C_0)A_{n+1} + k_2(_1C_1)A_n.$$

Similarly, one can put

$$A_{n+4} = k_1 A_{n+3} + k_2 A_{n+2}$$

$$= k_1[k_1(_1C_0)A_{n+2} + k_2(_1C_1)A_{n+1}] + k_2[k_1(_1C_0)A_{n+1} + k_2(_1C_1)A_n]$$

$$= k_1^2(_1C_0)A_{n+2} + k_1k_2(_1C_1)A_{n+1} + k_1 k_2(_1C_0)A_{n+1} + k_2^2(_1C_1)A_n$$

$$= k_1^2(_1C_0)A_{n+2} + k_1k_2(_1C_1 + _1C_0)A_{n+1} + k_2^2(_1C_1)A_n.$$

But $_1C_0 = 1 = _2C_0$, and $_1C_1 = 1 = _2C_2$, so this can be written as

$$A_{n+4} = k_1^2(_2C_0)A_{n+2} + k_1k_2(_1C_1 + _1C_0)A_{n+1} + k_2^2(_2C_2)A_n$$

$$= k_1^2(_2C_0)A_{n+2} + k_1k_2(_2C_1)A_{n+1} + k_2^2(_2C_2)A_n.$$

Following the same procedure for A_{n+6}, A_{n+8}, one might note that the first term of the expansion of A_{n+2q} always involves $_{q-1}C_0$, which is always 1 and thus equal to $_qC_0$, while the last term involves $_{q-1}C_{q-1}$, which is always 1 and thus equal to $_qC_q$, and every other term involves $(_{q-1}C_j + _{q-1}C_{j-1}) = _qC_j$, so that it can be seen that

$$A_{n+2q} = k_1^q(_qC_0)A_{n+q} + k_1^{q-1}k_2(_qC_1)A_{n+(q-1)} + k_1^{q-2}k_2^2(_qC_2)A_{n+(q-2)} + \dots$$

$$+ k_2^q(_qC_q)A_n.$$

For generalized Fibonacci sequences, where $k_1 = k_2 = 1$, this formula takes the form

$$G_{n+2q} = {}_qC_0 G_{n+q} + {}_qC_1 G_{n+(q-1)} + {}_qC_2 G_{n+(q-2)} + \ldots + {}_qC_q G_n.$$

This formula (sometimes written for generalized Fibonacci sequences, but frequently only given for the Fibonacci sequence) is found in a number of places, including Posamentier's book (see Bibliography), which illustrates it (as applied to the Fibonacci sequence) in some nice tables; here a similar table, but applied to a generalized Fibonacci sequence (whose first few terms are given in the bottom row of the table) instead, is given as Table 8:

q	0	1	2	3	4	5	6	7		
${}_0C_q$	1									
$\times G_q$	2									2
${}_1C_q$	1	1								
$\times G_q$	2	5								7
${}_2C_q$	1	2	1							
$\times G_q$	2	10	7							19
${}_3C_q$	1	3	3	1						
$\times G_q$	2	15	21	12						50
${}_4C_q$	1	4	6	4	1					
$\times G_q$	2	20	42	48	19					131
${}_5C_q$	1	5	10	10	5	1				
$\times G_q$	2	25	70	120	95	31				343
${}_6C_q$	1	6	15	20	15	6	1			
$\times G_q$	2	30	105	240	285	186	50			898
${}_7C_q$	1	7	21	35	35	21	7	1		
$\times G_q$	2	35	147	420	665	651	350	81		2351
G_q	2	5	7	12	19	31	50	81		

Table 8: Products of binomial coefficients with generalized Fibonacci sequence terms.

It can be seen that the rightmost column consists of every other term in the particular generalized Fibonacci sequence appearing in the bottom row of the table.

Obviously, by replacing n by $n - q$, it is seen that

$$A_{n+q} = k_1{}^q({}_qC_0)A_n + k_1{}^{q-1}k_2({}_qC_1)A_{n-1} + k_1{}^{q-2}k_2{}^2({}_qC_2)A_{n-2} + \ldots + k_2{}^q({}_qC_q)A_{n-q}$$

or

$$G_{n+q} = {}_qC_0 G_n + {}_qC_1 G_{n-1} + {}_qC_2 G_{n-2} + \ldots + {}_qC_q G_{n-q},$$

and by setting $n = 0$, it is seen that

$$A_{2q} = k_1{}^q({}_qC_0)A_q + k_1{}^{q-1}k_2({}_qC_1)A_{q-1} + k_1{}^{q-2}k_2{}^2({}_qC_2)A_{q-2} + \ldots + k_2{}^q({}_qC_q)A_0$$

or

$$G_{2q} = {}_qC_0G_q + {}_qC_1G_{q-1} + {}_qC_2G_{q-2} + \ldots + {}_qC_qG_0.$$

Since ${}_qC_p = {}_qC_{q-p}$, it can be seen that these formulas can also be written as

$$A_{n+2q} = k_1{}^q({}_qC_q)A_{n+q} + k_1{}^{q-1}k_2({}_qC_{q-1})A_{n+(q-1)} + k_1{}^{q-2}k_2{}^2({}_qC_{q-2})A_{n+(q-2)} + \ldots$$
$$+ k_2{}^q({}_qC_0)A_n,$$

$$G_{n+2q} = {}_qC_qG_{n+q} + {}_qC_{q-1}G_{n+(q-1)} + {}_qC_{q-2}G_{n+(q-2)} + \ldots + {}_qC_0G_n,$$

$$A_{n+q} = k_1{}^q({}_qC_q)A_n + k_1{}^{q-1}k_2({}_qC_{q-1})A_{n-1} + k_1{}^{q-2}k_2{}^2({}_qC_{q-2})A_{n-2} + \ldots$$
$$+ k_2{}^q({}_qC_0)A_{n-q},$$

$$G_{n+q} = {}_qC_qG_n + {}_qC_{q-1}G_{n-1} + {}_qC_{q-2}G_{n-2} + \ldots + {}_qC_0G_{n-q},$$

$$A_{2q} = k_1{}^q({}_qC_q)A_q + k_1{}^{q-1}k_2({}_qC_{q-1})A_{q-1} + k_1{}^{q-2}k_2{}^2({}_qC_{q-2})A_{q-2} + \ldots + k_2{}^q({}_qC_0)A_0,$$

and

$$G_{2q} = {}_qC_qG_q + {}_qC_{q-1}G_{q-1} + {}_qC_{q-2}G_{q-2} + \ldots + {}_qC_0G_0.$$

Returning to the original **recursion** formula,

$$A_n = k_1A_{n-1} + k_2A_{n-2}$$

can be written as

$$A_{n-2} = (-k_1/k_2)A_{n-1} + (1/k_2)A_n$$

which means that, replacing $n + i$ by $n - i$ wherever it occurs in the preceding argument, as well as replacing k_1 by $(-k_1/k_2)$ and k_2 by $(1/k_2)$, the six results

$$A_{n-2q} = (-k_1/k_2)^q({}_qC_0)A_{n-q} + (-k_1)^{q-1}({}_qC_1)A_{n-(q-1)}/k_2{}^q$$
$$+ (-k_1)^{q-2}({}_qC_2)A_{n-(q-2)}/k_2{}^q + \ldots + ({}_qC_q)A_n/k_2{}^q$$
$$= [(-k_1)^q({}_qC_0)A_{n-q} + (-k_1)^{q-1}({}_qC_1)A_{n-(q-1)} + (-k_1)^{q-2}({}_qC_2)A_{n-(q-2)}$$
$$+ \ldots + ({}_qC_q)A_n]/k_2{}^q,$$

$$A_{n-q} = (-k_1/k_2)^q({}_qC_0)A_n + (-k_1)^{q-1}({}_qC_1)A_{n+1}/k_2{}^q + (-k_1)^{q-2}({}_qC_2)A_{n+2}/k_2{}^q$$
$$+ \ldots + ({}_qC_q)A_{n+q}/k_2{}^q$$
$$= [(-k_1)^q({}_qC_0)A_n + (-k_1)^{q-1}({}_qC_1)A_{n+1} + (-k_1)^{q-2}({}_qC_2)A_{n+2} + \ldots$$
$$+ ({}_qC_q)A_{n+q}]/k_2{}^q,$$

$$A_{-2q} = (-k_1/k_2)^q({}_qC_0)A_{-q} + (-k_1)^{q-1}({}_qC_1)A_{-q-1}/k_2{}^q + (-k_1)^{q-2}({}_qC_2)A_{-q-2}/k_2{}^q$$
$$+ \ldots + ({}_qC_q)A_0/k_2{}^q$$
$$= [(-k_1)^q({}_qC_0)A_{-q} + (-k_1)^{q-1}({}_qC_1)A_{-q-1} + (-k_1)^{q-2}({}_qC_2)A_{-q-2}$$

$$+ \ldots + ({}_qC_q)A_0]/k_2{}^q,$$

$$A_{n-2q} = (-k_1/k_2)^q({}_qC_q)A_{n-q} + (-k_1)^{q-1}({}_qC_{q-1})A_{n-(q-1)}/k_2{}^q$$
$$+ (-k_1)^{q-2}({}_qC_{q-2})A_{n-(q-2)}/k_2{}^q + \ldots + ({}_qC_0)A_n/k_2{}^q$$
$$= [(-k_1)^q({}_qC_q)A_{n-q} + (-k_1)^{q-1}({}_qC_{q-1})A_{n-(q-1)} + (-k_1)^{q-2}({}_qC_{q-2})A_{n-(q-2)}$$
$$+ \ldots + ({}_qC_0)A_n]/k_2{}^q,$$

$$A_{n-q} = (-k_1/k_2)^q({}_qC_q)A_n + (-k_1)^{q-1}({}_qC_{q-1})A_{n+1}/k_2{}^q$$
$$+ (-k_1)^{q-2}({}_qC_{q-2})A_{n+2}/k_2{}^q + \ldots + ({}_qC_0)A_{n+q}/k_2{}^q$$
$$= [(-k_1)^q({}_qC_q)A_n + (-k_1)^{q-1}({}_qC_{q-1})A_{n+1} + (-k_1)^{q-2}({}_qC_{q-2})A_{n+2}$$
$$+ \ldots + ({}_qC_0)A_{n+q}]/k_2{}^q,$$

and

$$A_{-2q} = (-k_1/k_2)^q({}_qC_q)A_{-q} + (-k_1)^{q-1}({}_qC_{q-1})A_{-q-1}/k_2{}^q$$
$$+ (-k_1)^{q-2}({}_qC_{q-2})A_{-q-2}/k_2{}^q + \ldots + ({}_qC_0)A_0/k_2{}^q$$
$$= [(-k_1)^q({}_qC_q)A_{-q} + (-k_1)^{q-1}({}_qC_{q-1})A_{-q-1} + (-k_1)^{q-2}({}_qC_{q-2})A_{-q-2}$$
$$+ \ldots + ({}_qC_0)A_0]/k_2{}^q,$$

can be obtained. However, it is better to rearrange these formulas to give the terms in the opposite order, giving

$$A_{n-2q} = [{}_qC_qA_n - k_1({}_qC_{q-1})A_{n-1} + k_1{}^2({}_qC_{q-2})A_{n-2} -+ \ldots + (-k_1)^q({}_qC_0)A_{n-q}]/k_2{}^q,$$

$$A_{n-q} = [{}_qC_qA_{n+q} - k_1({}_qC_{q-1})A_{n+q-1} + k_1{}^2({}_qC_{q-2})A_{n+q-2} -+ \ldots$$
$$+ (-k_1)^q({}_qC_0)A_n]/k_2{}^q,$$

$$A_{-2q} = [{}_qC_qA_0 - k_1({}_qC_{q-1})A_{-1} + k_1{}^2({}_qC_{q-2})A_{-2} -+ \ldots$$
$$+ (-k_1)^q({}_qC_0)A_{-q}]/k_2{}^q,$$

$$A_{n-2q} = [{}_qC_0A_n - k_1({}_qC_1)A_{n-1} + k_1{}^2({}_qC_2)A_{n-2} -+ \ldots + (-k_1)^q({}_qC_q)A_{n-q}]/k_2{}^q,$$

$$A_{n-q} = [{}_qC_0A_{n+q} - k_1({}_qC_1)A_{n+q-1} + k_1{}^2({}_qC_2)A_{n+q-2} -+ \ldots$$
$$+ (-k_1)^q({}_qC_q)A_n]/k_2{}^q,$$

and

$$A_{-2q} = [{}_qC_0A_0 - k_1({}_qC_1)A_{-1} + k_1{}^2({}_qC_2)A_{-2} -+ \ldots + (-k_1)^q({}_qC_q)A_{-q}]/k_2{}^q.$$

For generalized Fibonacci sequences, it is only necessary to set $k_1 = k_2 = 1$, giving

$$G_{n-2q} = {}_qC_qG_n - {}_qC_{q-1}G_{n-1} + {}_qC_{q-2}G_{n-2} -+ \ldots + (-1)^q({}_qC_0)G_{n-q},$$

$$G_{n-q} = {}_qC_qG_{n+q} - {}_qC_{q-1}G_{n+q-1} + {}_qC_{q-2}G_{n+q-2} -+ \ldots$$
$$+ (-1)^q({}_qC_0)G_n,$$

$$G_{-2q} = {}_qC_qG_0 - {}_qC_{q-1}G_{-1} + {}_qC_{q-2}G_{-2} -+ \ldots + (-1)^q({}_qC_0)G_{-q},$$

$$G_{n-2q} = {}_qC_0G_n - {}_qC_1G_{n-1} + {}_qC_2G_{n-2} -+ \ldots + (-1)^q({}_qC_q)G_{n-q},$$

$$G_{n-q} = {}_qC_0G_{n+q} - {}_qC_1G_{n+q-1} + {}_qC_2G_{n+q-2} -+ \ldots + (-1)^q({}_qC_q)G_n,$$

and

$$G_{-2q} = {}_qC_0G_0 - {}_qC_1G_{-1} + {}_qC_2G_{-2} -+ \ldots + (-1)^q({}_qC_q)G_{-q}.$$

These formulas apply to all recurrent sequences A, and the special cases with $k_1 = k_2 = 1$ apply to all generalized Fibonacci sequences G, including the **Fibonacci** and Lucas sequences, so that they can be summarized:

$$A_{n+q} = k_1^q({}_qC_0)A_n + k_1^{q-1}k_2({}_qC_1)A_{q-1} + k_1^{q-2}k_2^2({}_qC_2)A_{q-2} + \ldots + k_2^q({}_qC_q)A_{n-q}$$

$$A_{n+q} = k_1^q({}_qC_q)A_n + k_1^{q-1}k_2({}_qC_{q-1})A_{q-1} + k_1^{q-2}k_2^2({}_qC_{q-2})A_{q-2} + \ldots$$
$$+ k_2^q({}_qC_0)A_{n-q}$$

$$G_{n+q} = {}_qC_0G_n + {}_qC_1G_{n-1} + {}_qC_2G_{n-2} + \ldots + {}_qC_qG_{n-q}$$

$$G_{n+q} = {}_qC_qG_n + {}_qC_{q-1}G_{n-1} + {}_qC_{q-2}G_{n-2} + \ldots + {}_qC_0G_{n-q}$$

$$F_{n+q} = {}_qC_0F_n + {}_qC_1F_{n-1} + {}_qC_2F_{n-2} + \ldots + {}_qC_qF_{n-q}$$

$$F_{n+q} = {}_qC_qF_n + {}_qC_{q-1}F_{n-1} + {}_qC_{q-2}F_{n-2} + \ldots + {}_qC_0F_{n-q}$$

$$L_{n+q} = {}_qC_0L_n + {}_qC_1L_{n-1} + {}_qC_2L_{n-2} + \ldots + {}_qC_qL_{n-q}$$

$$L_{n+q} = {}_qC_qL_n + {}_qC_{q-1}L_{n-1} + {}_qC_{q-2}L_{n-2} + \ldots + {}_qC_0L_{n-q}$$

$$A_{n-q} = [{}_qC_qA_{n+q} - k_1({}_qC_{q-1})A_{n+q-1} + k_1^2({}_qC_{q-2})A_{n+q-2} -+ \ldots$$
$$+ (-k_1)^q({}_qC_0)A_n]/k_2^q$$

$$A_{n-q} = [{}_qC_0A_{n+q} - k_1({}_qC_1)A_{n+q-1} + k_1^2({}_qC_2)A_{n+q-2} -+ \ldots + (-k_1)^q({}_qC_q)A_n]/k_2^q$$

$$G_{n-q} = {}_qC_0G_{n+q} - {}_qC_1G_{n+q-1} + {}_qC_2G_{n+q-2} -+ \ldots + (-1)^q{}_qC_qG_n$$

$$G_{n-q} = {}_qC_qG_{n+q} - {}_qC_{q-1}G_{n+q-1} + {}_qC_{q-2}G_{n+q-2} -+ \ldots + (-1)^q{}_qC_0G_n$$

$$F_{n-q} = {}_qC_0F_{n+q} - {}_qC_1F_{n+q-1} + {}_qC_2F_{n+q-2} -+ \ldots + (-1)^q{}_qC_qF_n$$

$$F_{n-q} = {}_qC_qF_{n+q} - {}_qC_{q-1}F_{n+q-1} + {}_qC_{q-2}F_{n+q-2} -+ \ldots + (-1)^q{}_qC_0F_n$$

$$L_{n-q} = {}_qC_0L_{n+q} - {}_qC_1L_{n+q-1} + {}_qC_2L_{n+q-2} -+ \ldots + (-1)^q{}_qC_qL_n$$

$$L_{n-q} = {}_qC_qL_{n+q} - {}_qC_{q-1}L_{n+q-1} + {}_qC_{q-2}L_{n+q-2} -+ \ldots + (-1)^q{}_qC_0L_n$$

$$A_{2q} = k_1^q({}_qC_0)A_q + k_1^{q-1}k_2({}_qC_1)A_{q-1} + k_1^{q-2}k_2^2({}_qC_2)A_{q-2} + \ldots + k_2^q({}_qC_q)A_0$$

$$A_{2q} = k_1^q({}_qC_q)A_q + k_1^{q-1}k_2({}_qC_{q-1})A_{q-1} + k_1^{q-2}k_2^2({}_qC_{q-2})A_{q-2} + \ldots + k_2^q({}_qC_0)A_0$$

$$G_{2q} = {}_qC_0G_q + {}_qC_1G_{q-1} + {}_qC_2G_{q-2} + \ldots + {}_qC_qG_0$$

$$G_{2q} = {}_qC_qG_q + {}_qC_{q-1}G_{q-1} + {}_qC_{q-2}G_{q-2} + \ldots + {}_qC_0G_0$$

$$F_{2q} = {}_qC_0F_q + {}_qC_1F_{q-1} + {}_qC_2F_{q-2} + \ldots + {}_qC_qF_0$$

$$F_{2q} = {}_qC_qF_q + {}_qC_{q-1}F_{q-1} + {}_qC_{q-2}F_{q-2} + \ldots + {}_qC_0F_0$$

$$L_{2q} = {}_qC_0L_q + {}_qC_1L_{q-1} + {}_qC_2L_{q-2} + \ldots + {}_qC_qL_0$$

$$L_{2q} = {}_qC_qL_q + {}_qC_{q-1}L_{q-1} + {}_qC_{q-2}L_{q-2} + \ldots + {}_qC_0L_0$$

$$A_{-2q} = [{}_qC_0A_0 - k_1({}_qC_1)A_{-1} + k_1^2({}_qC_2)A_{-2} - + \ldots + (-k_1)^q({}_qC_q)A_{-q}]/k_2^q$$

$$A_{-2q} = [{}_qC_qA_0 - k_1({}_qC_{q-1})A_{-1} + k_1^2({}_qC_{q-2})A_{-2} - + \ldots + (-k_1)^q({}_qC_0)A_{-q}]/k_2^q$$

$$G_{-2q} = {}_qC_0G_0 - {}_qC_1G_{-1} + {}_qC_2G_{-2} - + \ldots + (-1)^q{}_qC_qG_{-q}$$

$$G_{-2q} = {}_qC_qG_0 - {}_qC_{q-1}G_{-1} + {}_qC_{q-2}G_{-2} - + \ldots + (-1)^q{}_qC_0G_{-q}$$

$$F_{-2q} = {}_qC_0F_0 - {}_qC_1F_{-1} + {}_qC_2F_{-2} - + \ldots + (-1)^q{}_qC_qF_{-q}$$

$$F_{-2q} = {}_qC_qF_0 - {}_qC_{q-1}F_{-1} + {}_qC_{q-2}F_{-2} - + \ldots + (-1)^q{}_qC_0F_{-q}$$

$$L_{-2q} = {}_qC_0L_0 - {}_qC_1L_{-1} + {}_qC_2L_{-2} - + \ldots + (-1)^q{}_qC_qL_{-q}$$

$$L_{-2q} = {}_qC_qL_0 - {}_qC_{q-1}L_{-1} + {}_qC_{q-2}L_{-2} - + \ldots + (-1)^q{}_qC_0L_{-q}$$

$$A_{n+2q} = k_1^q({}_qC_0)A_{n+q} + k_1^{q-1}k_2({}_qC_1)A_{n+(q-1)} + k_1^{q-2}k_2^2({}_qC_2)A_{n+(q-2)} + \ldots$$
$$+ k_2^q({}_qC_q)A_n$$

$$A_{n+2q} = k_1^q({}_qC_q)A_{n+q} + k_1^{q-1}k_2({}_qC_{q-1})A_{n+(q-1)} + k_1^{q-2}k_2^2({}_qC_{q-2})A_{n+(q-2)} + \ldots$$
$$+ k_2^q({}_qC_0)A_n$$

$$G_{n+2q} = {}_qC_0G_{n+q} + {}_qC_1G_{n+(q-1)} + {}_qC_2G_{n+(q-2)} + \ldots + {}_qC_qG_n$$

$$G_{n+2q} = {}_qC_qG_{n+q} + {}_qC_{q-1}G_{n+(q-1)} + {}_qC_{q-2}G_{n+(q-2)} + \ldots + {}_qC_0G_n$$

$$F_{n+2q} = {}_qC_0F_{n+q} + {}_qC_1F_{n+(q-1)} + {}_qC_2F_{n+(q-2)} + \ldots + {}_qC_qF_n$$

$$F_{n+2q} = {}_qC_qF_{n+q} + {}_qC_{q-1}F_{n+(q-1)} + {}_qC_{q-2}F_{n+(q-2)} + \ldots + {}_qC_0F_n$$

$$L_{n+2q} = {}_qC_0L_{n+q} + {}_qC_1L_{n+(q-1)} + {}_qC_2L_{n+(q-2)} + \ldots + {}_qC_qL_n$$

$$L_{n+2q} = {}_qC_qL_{n+q} + {}_qC_{q-1}L_{n+(q-1)} + {}_qC_{q-2}L_{n+(q-2)} + \ldots + {}_qC_0L_n$$

$$A_{n-2q} = [{}_qC_0A_n - k_1({}_qC_1)A_{n-1} + k_1^2({}_qC_2)A_{n-2} - + \ldots + (-k_1)^q({}_qC_q)A_{n-q}]/k_2^q$$

$$A_{n-2q} = [{}_qC_qA_n - k_1({}_qC_{q-1})A_{n-1} + k_1^2({}_qC_{q-2})A_{n-2} - + \ldots + (-k_1)^q({}_qC_0)A_{n-q}]/k_2^q$$

$$G_{n-2q} = {}_qC_0G_n - {}_qC_1G_{n-1} + {}_qC_2G_{n-2} -+ \ldots + (-1)^q{}_qC_qG_{n-q}$$

$$G_{n-2q} = {}_qC_qG_n - {}_qC_{q-1}G_{n-1} + {}_qC_{q-2}G_{n-2} -+ \ldots + (-1)^q{}_qC_0G_{n-q}$$

$$F_{n-2q} = {}_qC_0F_n - {}_qC_1F_{n-1} + {}_qC_2F_{n-2} -+ \ldots + (-1)^q{}_qC_qF_{n-q}$$

$$F_{n-2q} = {}_qC_qF_n - {}_qC_{q-1}F_{n-1} + {}_qC_{q-2}F_{n-2} -+ \ldots + (-1)^q{}_qC_0F_{n-q}$$

$$L_{n-2q} = {}_qC_0L_n - {}_qC_1L_{n-1} + {}_qC_2L_{n-2} -+ \ldots + (-1)^q{}_qC_qL_{n-q}$$

$$L_{n-2q} = {}_qC_qL_n - {}_qC_{q-1}L_{n-1} + {}_qC_{q-2}L_{n-2} -+ \ldots + (-1)^q{}_qC_0L_{n-q}$$

For more formulas involving binomial coefficients, consult Dunlap's or Vajda's book. However, most of the formulas in those books are not derived, so they will have to be accepted without demonstration.

Chapter 14: Recurrence relations involving term-by-term products and powers of generalized Fibonacci sequences.

In Chapter 7, the term *generalized Fibonacci sequence n-tuplewise product sequence* was introduced. It can be noted that, in each case, given only the number of different generalized Fibonacci sequences which are multiplied term by term, just because the recurrence relation of the product sequence depends only on the recursion coefficients of the sequences being multiplied together, and all generalized Fibonacci sequences have the same recursion coefficients, all generalized Fibonacci sequence *n*-tuplewise product sequences with a given *n* have the same recursion coefficients. In Chapter 7 some equations were derived for a few of these recurrence relations involving term-by-term products of generalized Fibonacci sequences, which of course also apply to sequences consisting of powers of terms of generalized Fibonacci sequences. It will be easiest to think mainly in terms of power sequences, but it should be kept in mind that any formula involving a sequence consisting of the *n*th powers of terms of a generalized Fibonacci sequence can also be applied to *any* sequence consisting of term-by-term products of any n generalized Fibonacci sequences (*i. e.*, any generalized Fibonacci sequence *n*-tuplewise product sequence.) So if one has, just for example,

$$A_n = G_n{}^4 = k_1 G_{n-1}{}^4 + k_2 G_{n-2}{}^4 + k_3 G_{n-3}{}^4 + k_4 G_{n-4}{}^4 + k_5 G_{n-5}{}^4,$$

or in terms of *A* itself,

$$A_n = k_1 A_{n-1} + k_2 A_{n-2} + k_3 A_{n-3} + k_4 A_{n-4} + k_5 A_{n-5},$$

the same equation will apply if a sequence *A* is defined with

$$A_n = G_n G'_n G''_n G'''_n.$$

Many of the formulas in this chapter will be given without any derivation, because to do so would be extremely difficult and hard to follow. If one writes the generalized Fibonacci sequence definition formula in the **zero**-right-hand-side form

$$G_n - G_{n-1} - G_{n-2} = 0,$$

and the equations derived in Chapter 7 in the corresponding form

$$G_n{}^2 - 2G_{n-1}{}^2 - 2G_{n-2}{}^2 + G_{n-3}{}^2 = 0,$$

$$G_n{}^3 - 3G_{n-1}{}^3 - 6G_{n-2}{}^3 + 3G_{n-3}{}^3 + G_{n-4}{}^3 = 0, \text{ etc.},$$

and arranges the coefficients in a format resembling Pascal's triangle, one obtains Table 9 below.

n	p	0	1	2	3	4	5	6	7
		coefficient of G_{n-p}							
1		1	-1	-1					
2		1	-2	-2	1				
3		1	-3	-6	3	1			
4		1	-5	-15	15	5	-1		
5		1	-8	-40	60	40	-8	-1	
6		1	-13	-104	260	260	-104	13	1

Table 9: Recursion coefficients of powers of generalized Fibonacci sequences.

Some patterns can be observed:

1. Each column is either all positive or all negative. The first ($p = 0$) column is always positive, and then the columns run two negative, then two positive, etc. as far as needed.

2. The column for $p = 0$ is all 1's.

3. The magnitudes of the numbers in the column for $p = 1$ are simply the numbers in the Fibonacci sequence.

4. The magnitudes of the numbers in the column for $p = 2$ are the products of consecutive numbers in the Fibonacci sequence.

5. Each row is symmetrical (except for signs).

A formula can be used to compute these coefficients, somewhat similar to the formula for binomial coefficients, but involving products of terms in the Fibonacci sequence instead of factorials. However, it is easier to derive them using a formula that produces them recursively.

If $c_{n,p}$ represents the number appearing in Table 9 for the given value of n and p, once all the values for any n have been generated, the values for the next n are obtained by

$$c_{n+1,p} = \pm(F_{p+1}|c_{n,p}| + F_{n-p}|c_{n,p-1}|).$$

For an example, to calculate $c_{5,3}$, this formula with $n = 4$, $p = 3$ gives

$$c_{5,3} = F_4 c_{4,3} + F_1 c_{4,2}$$

$$= \pm[(3)(15) + (1)(15)]$$

giving the value 60, which has a positive sign because it is always + for $p = 1, 4, 5, 8, 9, ...$ and – for $p = 2, 3, 6, 7, ...$

Another interesting relationship is as follows. If any row is taken for the coefficients of a polynomial P_n (e. g., $P_3 = 1 - 3x - 6x^2 + 3x^3 + x^4$), the reciprocal of that polynomial, expressed as a power series, gives coefficients that form the column corresponding to that value of p of the table:

$$1/(1 - 3x - 6x^2 + 3x^3 + x^4) = 1 + 3x + 15x^2 + 60x^3 + 260\ x^4 + \dots$$

Chapter 15: Generation of recurrent sequences from binomial powers.

In Chapter 13, the focus was on the **binomial** coefficients $_nC_p$ and how they could be combined with terms of a particular generalized Fibonacci sequence to produce other terms in the same sequence. This chapter will also be based on the binomial theorem, but used in a different way to generate a recurrent sequence directly from the terms of a binomial expansion.

Logically, this chapter, which deals with sums of binomial coefficients, precedes the previous chapter, which deals with term-by-term sums of products involving the binomial coefficients with terms of recurrent sequences (including their special cases, generalized Fibonacci sequences and the **Fibonacci** and Lucas sequences); however, the order of chapters was reversed because the formulas of the previous chapter are easier to prove, and so more accessible than the formulas which will be demonstrated in this chapter.

Table 10, following, is basically a redoing of Table 6, but with the variables added in the form they would be for the expansion of powers of $(x + y)^n$.

p	0	1	2	3	4	5	6	7	8	9
n										
0	1									
1	x	y								
2	x^2	$2xy$	y^2							
3	x^3	$3x^2y$	$3xy^2$	y^3						
4	x^4	$4x^3y$	$6x^2y^2$	$4xy^3$	y^4					
5	x^5	$5x^4y$	$10x^3y^2$	$10x^2y^3$	$5xy^4$	y^5				
6	x^6	$6x^5y$	$15x^4y^2$	$20x^3y^3$	$15x^2y^4$	$6xy^5$	y^6			
7	x^7	$7x^6y$	$21x^5y^2$	$35x^4y^3$	$35x^3y^4$	$21x^2y^5$	$7xy^6$	y^7		
8	x^8	$8x^7y$	$28x^6y^2$	$56x^5y^3$	$70x^4y^4$	$56x^3y^5$	$28x^2y^6$	$8xy^7$	y^8	
9	x^9	$9x^8y$	$36x^7y^2$	$84x^6y^3$	$126x^5y^4$	$126x^4y^5$	$84x^3y^6$	$36x^2y^7$	$9xy^8$	y^9

Table 10: Powers of a binomial.

Normally, one looks at the rows of Table 10 and adds them up to form the powers of $(x + y)$. But now, consider the diagonal lines:

$$A_0 = 1,$$
$$A_1 = x,$$
$$A_2 = x^2 + y,$$
$$A_3 = x^3 + 2xy,$$
$$A_4 = x^4 + 3x^2y + y^2,$$

and so forth, in which A_{n+p} is defined as the sum of all the terms having a common value of $n + p$; the last such line that can be read off from this table being

$$A_9 = x^9 + 8x^7y + 21x^5y^2 + 20x^3y^3 + 5xy^4.$$
It is easy to see that for all $n > 0$ in these expressions,

$$A_{n+1} = xA_n + yA_{n-1}.$$

(This follows by considering each individual sum and the two preceding it, and substituting.) So if we put k_1 for x and k_2 for y, the A's form a recurrent sequence, with the k's being the recursion coefficients as they are normally written. In fact, as defined in Chapter 3, the A's are the prototypic co-recurrent sequence for any recurrent sequence with $k_1 = x$ and $k_2 = y$.

One special case of this occurs when $x = y = 1$. In that case, the binomial coefficients directly appear in Table 10, making it identical to Table 6. And the A_n values calculated from this procedure form the Fibonacci sequence (shifted by 1, so that $A_n = F_{n+1}$, as noted in Chapter 3). So, from binomial coefficients alone, the **Fibonacci** sequence can be generated! Suppose one takes all the binomial coefficients $_nC_p$ with a given sum $n + p$, and adds them. The result is shown below:

$n + p$	$_nC_p$ values		Sum
1	$_1C_0$	1	$1 = F_2$
2	$_2C_0 + {}_1C_1$	1+1	$2 = F_3$
3	$_3C_0 + {}_2C_1$	1+2	$3 = F_4$
4	$_4C_0 + {}_3C_1 + {}_2C_2$	1+3+1	$5 = F_5$
5	$_5C_0 + {}_4C_1 + {}_3C_2$	1+4+3	$8 = F_6$
6	$_6C_0 + {}_5C_1 + {}_4C_2 + {}_3C_3$	1+5+6+1	$13 = F_7$
7	$_7C_0 + {}_6C_1 + {}_5C_2 + {}_4C_3$	1+6+10+4	$21 = F_8$

Table 11: Generating the Fibonacci sequence from binomial coefficients.

It should be noticed that the number of terms goes up in jumps: 1, then 2 twice, then 3 twice, then 4 twice, and so on, governed by the fact that $p \le n$. Despite the irregularity, the Fibonacci sequence comes out precisely: the rightmost column is always F_{n+p+1}.

All the above applies only with reference to a sequence that is a prototypic co-recurrent sequence in the sense of Chapter 3. So how can one modify the Pascal triangle to produce *all* recurrent sequences by the same type of process?

Consider Table 12 below. It is simply a relabeling of Table 10; every cell in Table 12 is supposed to represent another way of writing the expression that appears in the same cell in Table 10. The symbol $c_{n,p}$ denotes the individual term found in the row for that value of n and the column for that value of p. And the actual expressions appearing in Table 10 were generated by the formulas:

$$c_{0,0} = 1,$$

$$c_{n,p} = k_1 c_{n-1,p} + k_2 c_{n-1,p-1}.$$

Here it is assumed that if $p = 0$, so that $p - 1 < 0$, the value 0 is substituted for $c_{n-1,p-1}$, and if $p = n$, so that $n - 1 < p$, the value 0 is substituted for $c_{n-1,p}$.

	p	0	1	2	3	4	5	6	7	8	9
n											
0		$c_{0,0}$									
1		$c_{1,0}$	$c_{1,1}$								
2		$c_{2,0}$	$c_{2,1}$	$c_{2,2}$							
3		$c_{3,0}$	$c_{3,1}$	$c_{3,2}$	$c_{3,3}$						
4		$c_{4,0}$	$c_{4,1}$	$c_{4,2}$	$c_{4,3}$	$c_{4,4}$					
5		$c_{5,0}$	$c_{5,1}$	$c_{5,2}$	$c_{5,3}$	$c_{5,4}$	$c_{5,5}$				
6		$c_{6,0}$	$c_{6,1}$	$c_{6,2}$	$c_{6,3}$	$c_{6,4}$	$c_{6,5}$	$c_{6,6}$			
7		$c_{7,0}$	$c_{7,1}$	$c_{7,2}$	$c_{7,3}$	$c_{7,4}$	$c_{7,5}$	$c_{7,6}$	$c_{7,7}$		
8		$c_{8,0}$	$c_{8,1}$	$c_{8,2}$	$c_{8,3}$	$c_{8,4}$	$c_{8,5}$	$c_{8,6}$	$c_{8,7}$	$c_{8,8}$	
9		$c_{9,0}$	$c_{9,1}$	$c_{9,2}$	$c_{9,3}$	$c_{9,4}$	$c_{9,5}$	$c_{9,6}$	$c_{9,7}$	$c_{9,8}$	$c_{9,9}$

Table 12: Relabeled version of the binomial power table.

When taking the diagonal sums generating the recurrent sequence A from the Pascal triangle in Table 10, the results were (in terms of the relabeling of Table 12):

$$A_0 = c_{0,0}$$

$$A_1 = c_{1,0}$$

$$A_2 = c_{2,0} + c_{1,1}$$

$$A_3 = c_{3,0} + c_{2,1},$$

etc., in which every $c_{n,p}$ term for which $n + p = q$ appears in the expression for A_q. One might note that the formula

$$c_{n,p} = k_1 c_{n-1,p} + k_2 c_{n-1,p-1}$$

means that every term $c_{n,p}$ for which $n + p = q$ (and which thus appears in the expression for A_q) is the sum of k_1 times a term which appears in the expression for A_{q-1} (namely $c_{n-1,p}$) and k_2 times a term which appears in the expression for A_{q-2} (namely $c_{n-1,p-1}$). So this demonstrates why this diagonal sum generates the recurrent sequence A. But in Table 10, the number 1 appeared for $c_{0,0}$, and x, which is equal to k_1, appeared for $c_{1,0}$. So *this*, therefore, is why the sequence A that was generated was prototypic. It is clear that if the values of A_0 and A_1 for *any recurrent sequence in question* are put for $c_{0,0}$ and $c_{1,0}$ in a table of the form of Table 12, the diagonal sum process of Table 11 will generate that sequence A.

An example of this would be to use this process to generate the Lucas sequence, which is not a prototypic co-recurrent sequence in the same way as the Fibonacci sequence, and the discussion above implies that, if one constructs a table like Table 6 but with 2 in the $c_{0,0}$ position, it will lead to a table in which all the 1's in the rightmost column are replaced by 2's, then (since the Lucas sequence is a generalized Fibonacci sequence so that $k_1 = k_2 = 1$) if the normal procedure where every number in the table is the sum of the one directly above it and the one above and one space to the left is followed, and then the same summing process as in Table 11 is used, the Lucas sequence will be obtained. This can be seen by looking at Tables 13 and 14 below. (In Table 14, the symbol C' will be used with the significance that $_nC'_p$ means the number appearing in the same position in Table 13 as the place where $_nC_p$ appears in Table 6.) One other difference is that, while Table 11 generated, for each given value of $n + p$, the value of F_{n+p+1}, Table 14 generates, instead, L_{n+p}.

	p	0	1	2	3	4	5	6	7	8	9
n											
0		2									
1		1	2								
2		1	3	2							
3		1	4	5	2						
4		1	5	9	7	2					
5		1	6	14	16	9	2				
6		1	7	20	30	25	11	2			
7		1	8	27	50	55	36	13	2		
8		1	9	35	77	105	91	49	15	2	
9		1	10	44	112	182	196	140	64	17	2

Table 13: Modified Pascal triangle which generates the Lucas sequence.

$n + p$	${}_nC'_p$ values		Sum
1	${}_1C'_0$	1	$1 = L_1$
2	${}_2C'_0 + {}_1C'_1$	1+2	$3 = L_2$
3	${}_3C'_0 + {}_2C'_1$	1+3	$4 = L_3$
4	${}_4C'_0 + {}_3C'_1 + {}_2C'_2$	1+4+2	$7 = L_4$
5	${}_5C'_0 + {}_4C'_1 + {}_3C'_2$	1+5+5	$11 = L_5$
6	${}_6C'_0 + {}_5C'_1 + {}_4C'_2 + {}_3C'_3$	1+6+9+2	$18 = L_6$
7	${}_7C'_0 + {}_6C'_1 + {}_5C'_2 + {}_4C'_3$	1+7+14+7	$29 = L_7$

Table 14: Use of modified Pascal triangle to generate the Lucas sequence.

Although it is not very clear from the arrangement in Table 13, the diagonal arrays of numbers with $n - p$ given certain numbers are interesting. All the numbers with $n - p = 0$ are 2's, the numbers with $n - p = 1$ are the odd numbers in order, and the numbers with $n - p = 2$ are the squares.

The procedure described here for generating a recurrent sequence from a modified Pascal triangle is hardly a practical way of producing the desired sequence; the recursion formula or the Binet-like formula will usually be much more convenient. (In fact, generating the elements in the modified Pascal triangle really involves applying the recursion formula, but more times than generating the sequence directly from the recursion formula would require!) Rather, the intent in this chapter is to call the reader's attention to the patterns relating binomial power expansions to recurrent sequences. And it also should remedy one problem. Most books, including such excellent ones as the ones by Dunlap and Vajda in the bibliography, treat the formula implied by the procedure of Table 11 for the Fibonacci sequence as a magic formula that is specialized for that sequence. In Posamentier's book, the procedure given in this chapter for the Lucas sequence (*i. e.,* Table 14) also appears, but since that book does not even *refer* to generalized Fibonacci sequences (not to mention recurrent sequences in general!) it is still not mentioned there that the procedures of Tables 11 and 14 are special cases of the more general relationships derived in this chapter. But in fact, this formula cited in all the books in question simply works for the Fibonacci sequence because $F_1 = F_2 = k_1 = k_2 = 1$. And so the general process reduces to the procedure of Table 11.

Chapter 16: The difference-of-products formula introduced.

The next four chapters will use the concept of co-recurrent sequences extensively, first deriving and then applying a formula which, in most books, is only shown for the special case of the Fibonacci sequence, or perhaps the Lucas sequence as well. Many people who are reading this book may already have seen the Fibonacci sequence, but are not familiar with the general concept of a recurrent sequence, of which the Fibonacci sequence is just *one special case*. And similarly, one of the best-known properties of the Fibonacci sequence is that the square of any term always differs by one from the product of the two on either side of it, or expressed algebraically,

$$F_n^2 = F_{n-1}F_{n+1} \pm 1,$$

but hardly anyone is aware that *all recurrent sequences* have a property which reduces to that equation as a special case. The formula is one that is not usually given a name, but it will be useful to have one: it will be referred to as the **difference-of-products formula** in this book.

The difference-of-products formula is actually one of the most important properties of recurrent sequences, which is why four whole chapters of this book are devoted to it. As just mentioned in the preceding paragraph, the difference between the approach in this book and the way it is usually treated is that in this book, the difference-of-products formula is shown to apply to recurrent sequences in general, and its special cases for the Fibonacci and Lucas sequences are simply shown to be just that: special cases of a more general difference-of-products formula, as opposed to the usual treatment, where these special cases are treated in isolation.

Back in Chapter 3, the concept of the *prototypic co-recurrent sequence* was introduced. The definition should be reviewed now if it has not yet been learned, as the difference-of-products formula requires the concept in its statement. As was stated in Chapter 3, the Fibonacci sequence is the prototypic co-recurrent sequence of any generalized Fibonacci sequence, a property that causes it to occur in many of the special cases of the difference-of-products formula (as will be seen in Chapter 18).

The difference-of-products formula applies to corresponding terms in two co-recurrent sequences (*i. e.*, two terms in the same positions of both sequences). One form of the difference-of-products formula (others will be given in Chapter 18) states that, given any two sequences A and A', sharing the recursion formula

$$A_i = k_1 A_{i-1} + k_2 A_{i-2},$$

if A'' is the prototypic co-recurrent sequence of both A and A', then for any two terms A_p and A_q in one sequence and A'_p and A'_q in the other sequence,

$$A_p A'_q - A_q A'_p = (-k_2)^p A''_{q-p-1}(A_0 A'_1 - A_1 A'_0).$$

Chapter 17, which follows (and can be skipped if the reader wants to take it on faith) is the derivation of the difference-of-products formula, while Chapters 18 and 19 will be concerned with showing a number of alternative forms and special cases of the difference-of-products formula by choosing specific values of p, q, A, and/or A'.

To see that the formula $F_n^2 = F_{n-1}F_{n+1} \pm 1$, claimed to be a special case of the difference-of-products formula, follows from the formula as stated above, note that if A is the Fibonacci sequence, A' is the Fibonacci sequence shifted by 1 so that $A'_{n+1} = A_n$, and the values $p = n$, $q = n - 1$, and $k_2 = 1$ are taken (k_1 does not occur in the difference-of-products formula), the difference-of-products formula yields this equation $F_n^2 = F_{n-1}F_{n+1} \pm 1$. Chapter 19, as mentioned earlier, is concerned with the numerous special cases of the difference-of-products formula that can be derived by assuming specific values for the variables in that formula.

Chapter 17:   The derivation of the difference-of-products formula.

This chapter will be devoted to the derivation of a formula for the expression

$$A_p A'_q - A_q A'_p$$

where A and A' represent any two co-recurrent sequences. It is a long and involved derivation, and the reader may choose to take it on faith rather than follow, but it has been included because those readers who are more mathematically sophisticated will not be willing to take the difference-of-products formula on faith. Because it is so long and involved, a complete chapter has been devoted to this derivation. As the sequences A and A' are co-recurrent, they share a **recursion** formula, and the symbols k_1 and k_2 will be defined such that

$$A_i = k_1 A_{i-1} + k_2 A_{i-2},$$

with exactly the same coefficients k_1 and k_2 appearing in the expression for A'.

It is also necessary to consider the prototypic co-recurrent sequence A'', defined (as stated in Chapter 3) by $A''_0 = 1$, $A''_1 = k_1$, and $A''_i = k_1 A''_{i-1} + k_2 A''_{i-2}$. Before deriving the difference-of-products formula itself, it will be useful to come up with an explicit formula for the terms of this sequence. To this end, it should be recalled that

$$A''_i = a''_1 r^i + a''_2 r'^i, \text{ with}$$

$$A''_0 = 1 \text{ and } A''_1 = k_1.$$

In Chapter 6, a formula for the a's was explicitly derived, and putting in the double primes for this specific case:

$$a''_1 = (A''_1 - A''_0 r')/(r - r'),$$

$$a''_2 = (A''_0 r - A''_1)/(r - r'),$$

or, with the specific values $A''_0 = 1$ and $A''_1 = k_1$ inserted,

$$a''_1 = (k_1 - r')/(r - r') = (r' - k_1)/(r' - r),$$

$$a''_2 = (r - k_1)/(r - r') = (k_1 - r)/(r' - r).$$

Substituting,

$$A''_n = a''_1 r^i + a''_2 r'^i$$

$$= (r' - k_1) r^n/(r' - r) + (k_1 - r) r'^n/(r' - r)$$

But $k_1 = r + r'$, so $r' - k_1 = r$ and $r' - k_1 = r$, simplifying the above equation to:

$$A''_n = (r^{n+1} + r'^{n+1})/(r' - r).$$

This will be used later in the chapter. To get back to the difference-of-products formula, one goes back to the recurrent sequence to double-geometric sequence formulas, and since all three of the sequences A, A', and A'' share a common recursion formula with the coefficients k_1 and k_2, the double geometric sequences equivalent to A, A', and A'' share **common** ratios r and r', where

$$r = [k_1 + \sqrt{(k_1^2 + 4k_2)}]/2, \text{ and}$$

$$r' = [k_1 - \sqrt{(k_1^2 + 4k_2)}]/2.$$

The double geometric sequences are then given by

$$A_i = a_1 r^i + a_2 r'^i,$$

$$A'_i = a'_1 r^i + a'_2 r'^i,$$

$$A''_i = a''_1 r^i + a''_2 r'^i.$$

For the moment, the a's will not be calculated, as it is easier to work with them in the form given here. Substituting,

$$A_p A'_q - A_q A'_p$$

$$= (a_1 r^p + a_2 r'^p)(a'_1 r^q + a'_2 r'^q) - (a_1 r^q + a_2 r'^q)(a'_1 r^p + a'_2 r'^p)$$

$$= (a_1 r^p a'_1 r^q + a_1 r^p a'_2 r'^q + a_2 r'^p a'_1 r^q + a_2 r'^p a'_2 r'^q) -$$
$$(a_1 r^q a'_1 r^p + a_1 r^q a'_2 r'^p + a_2 r'^q a'_1 r^p + a_2 r'^q a'_2 r'^p)$$

$$= a_1 r^p a'_1 r^q + a_1 r^p a'_2 r'^q + a_2 r'^p a'_1 r^q + a_2 r'^p a'_2 r'^q -$$
$$a_1 r^q a'_1 r^p - a_1 r^q a'_2 r'^p - a_2 r'^q a'_1 r^p - a_2 r'^q a'_2 r'^p$$

$$= a_1 a'_1 r^p r^q + a_1 a'_2 r^p r'^q + a_2 a'_1 r'^p r^q + a_2 a'_2 r'^p r'^q -$$
$$a_1 a'_1 r^q r^p - a_1 a'_2 r^q r'^p - a_2 a'_1 r'^q r^p - a_2 a'_2 r'^q r'^p$$

$$= a_1 a'_2 r^p r'^q + a_2 a'_1 r'^p r^q - a_1 a'_2 r^q r'^p - a_2 a'_1 r'^q r^p$$
$$= a_1 a'_2 r^p r'^q - a_1 a'_2 r^q r'^p + a_2 a'_1 r'^p r^q - a_2 a'_1 r'^q r^p$$
$$= a_1 a'_2 (r^p r'^q - r^q r'^p) + a_2 a'_1 (r'^p r^q - r^p r'^q)$$
$$= (a_1 a'_2 - a_2 a'_1)(r^p r'^q - r^q r'^p)$$
$$= (a_1 a'_2 - a_2 a'_1) r^p r'^p (r'^{q-p} - r^{q-p}).$$

By exactly the same process, so all the intermediate steps can be omitted,

$$A_0 A'_1 - A_1 A'_0 = (a_1 a'_2 - a_2 a'_1)(r^0 r'^1 - r'^0 r^1)$$

$$= (a_1 a'_2 - a_2 a'_1)(r' - r).$$

So the exact expression of the factor $(a_1 a'_2 - a_2 a'_1)$ in terms of any of the parameters involved can be ignored, as it is the same in both expressions, and one can simply put

$$a_1a'_2 - a_2a'_1 = (A_0A'_1 - A_1A'_0)/(r' - r)$$

into the equation

$$A_pA'_q - A_qA'_p = (a_1a'_2 - a_2a'_1)r^p r'^p(r'^{q-p} - r^{q-p})$$

giving

$$A_pA'_q - A_qA'_p = r^p r'^p(r'^{q-p} - r^{q-p})(A_0A'_1 - A_1A'_0)/(r' - r).$$

This is to be compared with the difference-of-products formula, given in Chapter 16:

$$A_pA'_q - A_qA'_p = (-k_2)^p A''_{q-p-1}(A_0A'_1 - A_1A'_0).$$

It has already been demonstrated (Chapter 4) that

$$k_2 = -rr',$$

so that $r^p r'^p$ can be written as $(-k_2)^p$, and as the difference-of-products formula contains the factor $(A_0A'_1 - A_1A'_0)$, all that is necessary at this point is to show that

$$A''_{q-p-1} = (r'^{q-p} - r^{q-p})/(r' - r)$$

But earlier in this chapter it was established that

$$A''_n = (r^{n+1} + r'^{n+1})/(r' - r).$$

Replacing n by $q - p - 1$, this gives the necessary relationship, and completes the derivation of the difference-of-products formula.

Chapter 18: Alternative forms of the difference-of-products formula.

Earlier, in Chapter 16, the *difference-of-products formula* was defined as

$$A_pA'_q - A_qA'_p = (-k_2)^p A''_{q-p-1}(A_0A'_1 - A_1A'_0).$$

It can be generalized to produce an even more powerful equation. For if one writes

$$A_mA'_n - A_nA'_m = (-k_2)^m A''_{n-m-1}(A_0A'_1 - A_1A'_0) \text{ and}$$
$$A_pA'_q - A_qA'_p = (-k_2)^p A''_{q-p-1}(A_0A'_1 - A_1A'_0),$$

they can be combined to eliminate the common factor $(A_0A'_1 - A_1A'_0)$:

$$(A_mA'_n - A_nA'_m)/(A_pA'_q - A_qA'_p) = (-k_2)^{m-p}(A''_{n-m-1}/A''_{q-p-1}),$$

where, as before, A and A' are any two co-recurrent sequences and A'' represents the prototypic co-recurrent sequence common to A and A'. This, rather than

$$A_pA'_q - A_qA'_p = (-k_2)^p A''_{q-p-1}(A_0A'_1 - A_1A'_0),$$

could be taken as *the* difference-of-products formula, and in order to distinguish them in the rest of this discussion, the name **general difference-of-products formula** will refer to

$$(A_mA'_n - A_nA'_m)/(A_pA'_q - A_qA'_p) = (-k_2)^{m-p}(A''_{n-m-1}/A''_{q-p-1}),$$

while the name **simple difference-of-products formula** will refer to

$$A_pA'_q - A_qA'_p = (-k_2)^p A''_{q-p-1}(A_0A'_1 - A_1A'_0).$$

For generalized Fibonacci sequences, as was stated earlier, the prototypic co-recurrent sequence is the Fibonacci sequence (but shifted by 1: $A''_j = F_{j+1}$), and the **general difference-of-products formula** takes the form

$$(G_mG'_n - G_nG'_m)/(G_pG'_q - G_qG'_p) = (-1)^{m-p}(F_{n-m}/F_{q-p}),$$

while the **simple difference-of-products formula** takes the form

$$G_pG'_q - G_qG'_p = (-1)^p F_{q-p}(G_0G'_1 - G_1G'_0).$$

In the simple difference-of-products formula, the terms of the sequence with the subscripts 0 and 1 are specifically highlighted. The reader may feel uncomfortable with this; you might prefer, for example, to express $A_pA'_q - A_qA'_p$ in terms of $A_1A'_2 - A_2A'_1$, for example. It is obvious that this can be done (in part, because any shifted form of a recurrent sequence is itself a recurrent sequence, co-recurrent with it!)

But rather than explicitly write all these forms, it can simply be noted that the general difference-of-products formula allows this: simply let m and n be any two fixed numbers (they do not even have to be consecutive!) and solve the resulting equation for $A_p A'_q - A_q A'_p$ in terms of $A_m A'_n - A_n A'_m$. The resulting forms will not be listed here, as they are easy to write down.

A number of special cases of these formulas are worthy of being explicitly stated. If A' is simply a shifted form of A, with the amount of shift being given by s, the two formulas take the form

$$(A_m A_{n+s} - A_n A_{m+s})/(A_p A_{q+s} - A_q A_{p+s}) = (-k_2)^{m-p}(A''_{n-m-1}/A''_{q-p-1}),$$

and

$$A_p A_{q+s} - A_q A_{p+s} = (-k_2)^p A''_{q-p-1}(A_0 A_{s+1} - A_1 A_s).$$

Replacing m, n, p, and q by m, $m+i$, p, and $p+j$, they become

$$(A_m A'_{m+i} - A_{m+i} A'_m)/(A_p A'_{p+j} - A_{p+j} A'_p) = (-k_2)^{m-p}(A''_{i-1}/A''_{j-1}),$$

$$A_p A'_{p+j} - A_{p+j} A'_p = (-k_2)^p A''_{j-1}(A_0 A'_1 - A_1 A'_0),$$

$$(A_m A_{m+i+s} - A_{m+i} A_{m+s})/(A_p A_{p+j+s} - A_{p+j} A_{p+s}) = (-k_2)^{m-p}(A''_{i-1}/A''_{j-1}),$$

and

$$A_p A_{p+j+s} - A_{p+j} A_{p+s} = (-k_2)^p A''_{j-1}(A_0 A_{s+1} - A_1 A_s).$$

For the specific case of generalized Fibonacci sequences, these formulas become

$$(G_m G_{n+s} - G_n G_{m+s})/(G_p G_{q+s} - G_q G_{p+s}) = (-1)^{m-p}(F_{n-m}/F_{q-p}),$$

$$G_p G_{q+s} - G_q G_{p+s} = (-1)^p F_{q-p}(G_0 G_{s+1} - G_1 G_s),$$

$$(G_m G'_{m+i} - G_{m+i} G'_m)/(G_p G'_{p+j} - G_{p+j} G'_p) = (-1)^{m-p}(F_i/F_j),$$

$$G_p G'_{p+j} - G_{p+j} G'_p = (-1)^p F_j(G_0 G'_1 - G_1 G'_0),$$

$$(G_m G_{m+i+s} - G_{m+i} G_{m+s})/(G_p G_{p+j+s} - G_{p+j} G_{p+s}) = (-1)^{m-p}(F_i/F_j),$$

and

$$G_p G_{p+j+s} - G_{p+j} G_{p+s} = (-1)^p F_j(G_0 G_{s+1} - G_1 G_s).$$

Naturally, when G or G' is specifically the Fibonacci sequence F or the Lucas sequence L, the values $F_0 = 0$, $F_1 = 1$, $L_0 = 2$, and $L_1 = 1$ must be put in.

Chapter 19: Additional difference-of-products expressions.

When the two sequences appearing in the difference-of-products formula are both *generalized Fibonacci sequences* such that one is a **shifted** form of the other, some particularly symmetric formulas can be generated. For example, it was shown that

$$G_pG'_{p+j} - G_{p+j}G'_p = (-1)^pF_j(G_0G'_1 - G_1G'_0).$$

But if $G'_p = G_{p+j}$, this becomes

$$G_pG_{p+2j} - G_{p+j}^2 = (-1)^pF_j(G_0G_{1+j} - G_1G_j).$$

This equation can be simplified, though this requires the use of the formula

$$F_{n-1}G_0 + F_nG_1 = G_n$$

which has not yet been demonstrated (it will be in Chapter 20). With that formula, it is possible to write

$$G_0G_{1+j} - G_1G_j = G_0(F_jG_0 + F_{1+j}G_1) - G_1(F_{j-1}G_0 + F_jG_1)$$

$$= (F_jG_0^2 + F_{1+j}G_0G_1) - (F_{j-1}G_0G_1 + F_jG_1^2)$$

$$= F_jG_0^2 + F_{1+j}G_0G_1 - F_{j-1}G_0G_1 - F_jG_1^2$$

$$= F_jG_0^2 - F_jG_1^2 + F_{1+j}G_0G_1 - F_{j-1}G_0G_1$$

$$= F_jG_0^2 - F_jG_1^2 + (F_{j+1} - F_{j-1})G_0G_1$$

$$= F_jG_0^2 - F_jG_1^2 + F_jG_0G_1$$

$$= F_j(G_0^2 - G_1^2 + G_0G_1)$$

$$= F_j(G_0^2 + G_0G_1 - G_1^2)$$

$$= F_j[G_0(G_0 + G_1) - G_1^2]$$

$$= F_j(G_0G_2 - G_1^2).$$

So, substituting in the equation for $G_pG_{p+2j} - G_{p+j}^2$ above, it can be written

$$G_pG_{p+2j} - G_{p+j}^2 = (-1)^pF_j^2(G_0G_2 - G_1^2).$$

What this means is that if you take *any three* terms in a generalized Fibonacci sequence, such that the spacing between the first and the middle is the same as the spacing between the middle and the last, square the middle one and multiply the first and last together, the answer will be the same (except for sign) as if you take any other three terms with the same spacing. It might be noted that the expression $G_0G_2 - G_1^2$ is characteristic of the sequence and does not depend on p or j; also that

$$F_0 F_2 - F_1^2 = (0)(1) - 1 = -1,$$

and

$$L_0 L_2 - L_1^2 = (2)(3) - 1 = 5.$$

Tables 15 and 16 illustrate this for the sequence 3, 2, 5, 7, ..., for example.

G_p	G_{p+j}	G_{p+2j}	$G_{p+j}^2 - G_p G_{p+2j}$
3	7	31	$(3)^2 - (7)(31) = -44$
2	12	50	$(2)^2 - (12)(50) = 44$
5	19	81	$(5)^2 - (19)(81) = -44$
7	31	131	$(7)^2 - (31)(131) = 44$
12	50	212	-44
19	81	343	44
31	131	555	-44
50	212	898	44
81	343	1453	-44
131	555	2351	44
212	898	3804	-44
343	1453	6155	44
555	2351	9959	-44
898	3804	16114	44

Table 15: Table showing values of $G_{p+j}^2 - G_p G_{p+2j}$ for $j=3$.

Note that in Table 15, the rightmost column alternates between plus and minus 44, where $44 = 2^2(11)$, and $2 = F_3$, as demanded by the formula $G_p G_{p+2j} - G_{p+j}^2 = (-1)^p F_j^2 (G_0 G_2 - G_1^2)$, while in Table 16, the rightmost column alternates between plus and minus 275, where $275 = 5^2(11)$, and $5 = F_5$.

G_p	G_{p+j}	G_{p+2j}	$G_{p+j}^2 - G_p G_{p+2j}$
3	19	212	-275
2	31	343	275
5	50	555	-275
7	81	898	275
12	131	1453	-275
19	212	2351	275
31	343	3804	-275
50	555	6155	275
81	898	9959	-275
131	1453	16114	275

212	2351	26073	-275
343	3804	42187	275
555	6155	68260	-275
898	9959	110447	275

Table 16: Table showing values of $G_{p+j}^2 - G_pG_{p+2j}$ for $j=5$.

The appearance of F_j^2 in the formula gives rise to Table 17; the factor of 11 simply comes from the term $G_0G_2 - G_1^2$.

j	G_p	G_{p+j}	G_{p+2j}	$G_{p+j}^2 - G_pG_{p+2j}$
1		2	5	-11
2		5	12	-11
3		7	31	-44
4	3	12	81	-99
5		19	212	-275
6		31	555	-704
7		50	1453	-1859

Table 17: Table showing values of $G_{p+j}^2 - G_pG_{p+2j}$ for varying j.

As was done for the expression $G_{p+j}^2 - G_pG_{p+2j}$, a simple (and rather similar) form can also be derived for the expression $G_{p+j}G_{p+j+1} - G_pG_{p+2j+1}$, which is also a symmetric arrangement. One starts with the same equation:

$$G_pG'_{p+j} - G_{p+j}G'_p = (-1)^pF_j(G_0G'_1 - G_1G'_0).$$

And setting $G'_p = G_{p+j+1}$ (as before G'_p was set to G_{p+j}), this equation becomes

$$G_pG_{p+2j+1} - G_{p+j}G_{p+j+1} = (-1)^pF_j(G_0G_{j+2} - G_1G_{j+1}).$$

Since it has been demonstrated that

$$G_0G_{1+j} - G_1G_j = F_j(G_0G_2 - G_1^2)$$

for all j, one can simply replace j by $j + 1$ to obtain

$$G_0G_{j+2} - G_1G_{j+1} = F_{j+1}(G_0G_2 - G_1^2).$$

Thus,

$$G_pG_{p+2j+1} - G_{p+j}G_{p+j+1} = (-1)^pF_jF_{j+1}(G_0G_2 - G_1^2).$$

Note that except for having F_jF_{j+1} instead of F_j^2, this equation is the same as the one for $G_pG_{p+2j} - G_{p+j}G_{p+j}$.

Actually, though the mathematics is a bit more difficult, when A is *any* 3-term recurrent sequence, similar formulas for

$$A_pA_{p+2j} - A_{p+j}A_{p+j} \text{ and } A_pA_{p+2j+1} - A_{p+j}A_{p+j+1}$$

can be obtained. They do, not surprisingly, involve the k's and are otherwise identical to the ones derived for

$$G_p G_{p+2j} - G_{p+j} G_{p+j} \text{ and } G_p G_{p+2j+1} - G_{p+j} G_{p+j+1}.$$

Chapter 20:   The sum-of-products formulas.

In the last two chapters, expressions of the form

$$A_p A'_q - A_r A'_s$$

were considered, with specific choices, in some cases, of the sequences A and A' and the subscripts p, q, r, and s. This chapter is concerned with expressions of the form

$$A_p A'_q + A_r A''_s.$$

First consider a recurrent sequence A, which begins as follows:

$$A_0,$$

$$A_1,$$

$$k_2 A_0 + k_1 A_1,$$

$$k_2 A_1 + k_1 (k_2 A_0 + k_1 A_1),$$

$$k_2 (k_2 A_0 + k_1 A_1) + k_1 [k_2 A_1 + k_1 (k_2 A_0 + k_1 A_1)].$$

It will be useful to write all the terms in the form $u A_0 + v A_1$, as follows:

$$1(A_0) + 0(A_1),$$

$$0(A_0) + 1(A_1),$$

$$k_2 A_0 + k_1 A_1,$$

$$k_1 k_2 A_0 + (k_2 + k_1^2) A_1,$$

$$(k_2^2 + k_1^2 k_2) A_0 + (2 k_2 k_1 + k_1^3) A_1.$$

Clearly, if one takes A'_n and A''_n as the standard co-recurrent basis sequences of A, by setting $A'_0 = 1$, $A'_1 = 0$, $A''_0 = 0$, and $A''_1 = 1$, these can be written $A_n = A'_n A_0 + A''_n A_1$, and if the procedure is continued, it can be seen to follow forever. For

$$A_n = k_1 A_{n-1} + k_2 A_{n-2}$$

$$= k_1 (A'_{n-1} A_0 + A''_{n-1} A_1) + k_2 (A'_{n-2} A_0 + A''_{n-2} A_1)$$

$$= k_1 A'_{n-1} A_0 + k_1 A''_{n-1} A_1 + k_2 A'_{n-2} A_0 + k_2 A''_{n-2} A_1$$

$$= k_1 A'_{n-1} A_0 + k_2 A'_{n-2} A_0 + k_1 A''_{n-1} A_1 + k_2 A''_{n-2} A_1$$

$$= (k_1 A'_{n-1} + k_2 A'_{n-2}) A_0 + (k_1 A''_{n-1} + k_2 A''_{n-2}) A_1$$

$$= A'_n A_0 + A''_n A_1.$$

In conclusion, for any recurrent sequence A, if A' and A'' are the standard co-recurrent basis sequences of A, then

$$A'_n A_0 + A''_n A_1 = A_n$$

for all n. Obviously, although this discussion concerned individual *terms* of the sequence, the fact that A_0 and A_1 are constant for any individual **recurrent** sequence A means that one could consider this formula to be the expression of the recurrent sequence A in terms of its standard co-recurrent basis sequences:

$$A = A' A_0 + A'' A_1,$$

which, in the context of a vector space picture of the set of co-recurrent sequences, can be viewed as the resolution of A into components, for those readers who have some acquaintance with linear algebra. (For those who do not, this statement can be ignored, except that it explains why the term "**basis**" was used.)

And obviously, since any **shifted** form of A is also a co-recurrent sequence with A,

$$A'_n A_m + A''_n A_{m+1} = A_{m+n}$$

for all m and n. Now specifically when A is a *generalized Fibonacci sequence*, both the k's become 1, and this makes A' a shifted Fibonacci sequence with $A''_n = F_{n-1}$, while A'' is the Fibonacci sequence, so the two formulas become

$$F_{n-1} G_0 + F_n G_1 = G_n$$

and

$$F_{n-1} G_m + F_n G_{m+1} = G_{m+n}.$$

Note that if one takes $m = n - 1$ in the preceding equation, then increases all the subscripts by 1, one gets

$$F_n G_n + F_{n+1} G_{n+1} = G_{2n+1},$$

and if $G = F$, this in turn gives

$$F_n^2 + F_{n+1}^2 = F_{2n+1},$$

which says that the sum of the squares of *any two consecutive terms* of the Fibonacci sequence is itself a member of the Fibonacci sequence.

Of course, G here can be F, L, or a geometric progression with **common** ratio τ or τ'. However, if G is F, the equation

$$F_{n-1} G_0 + F_n G_1 = G_n$$

simply becomes

$$F_{n-1}(0) + F_n(1) = F_n,$$

which is trivial. The corresponding form when G is L is

$$F_{n-1}(2) + F_n(1) = L_n,$$

which is the useful formula

$$2F_{n-1} + F_n = L_n,$$

and even though the left-hand side of this equation does not look like a sum of products, it is included here, because it is in fact a special case of the sum-of-products formula given above.

Another sum-of-products formula is also useful. Consider four consecutive elements of G, namely

$$G_{m+1}, G_{m+2}, G_{m+3}, \text{ and } G_{m+4}.$$

Also consider four consecutive elements of G', namely

$$G'_{n+1}, G'_{n+2}, G'_{n+3}, \text{ and } G'_{n+4}.$$

Now, by the generalized Fibonacci sequence definition formula,

$$G_{m+3} = G_{m+1} + G_{m+2},$$
$$G_{m+4} = G_{m+1} + 2G_{m+2},$$
$$G'_{n+3} = G'_{n+1} + G'_{n+2}, \text{ and}$$
$$G'_{n+4} = G'_{n+1} + 2G'_{n+2}.$$

So

$$G_{m+3}G'_{n+4} = (G_{m+1} + G_{m+2})(G'_{n+1} + 2G'_{n+2}), \text{ and}$$

$$G'_{n+3}G_{m+4} = (G'_{n+1} + G'_{n+2})(G_{m+1} + 2G_{m+2}),$$

which means that

$$G_{m+3}G'_{n+4} + G'_{n+3}G_{m+4} = (G_{m+1} + G_{m+2})(G'_{n+1} + 2G'_{n+2})$$
$$+ (G'_{n+1} + G'_{n+2})(G_{m+1} + 2G_{m+2})$$

$$= (G_{m+1}G'_{n+1} + 2G_{m+1}G'_{n+2} + G_{m+2}G'_{n+1} + 2G_{m+2}G'_{n+2})$$
$$+ (G_{m+1}G'_{n+1} + 2G_{m+2}G'_{n+1} + G_{m+1}G'_{n+2} + 2G_{m+2}G'_{n+2})$$

$$= (2G_{m+1}G'_{n+1} + 3G_{m+1}G'_{n+2} + 3G_{m+2}G'_{n+1} + 4G_{m+2}G'_{n+2}).$$

But also,

$$G_{m+2}G'_{n+2} + G_{m+3}G'_{n+3} = G_{m+2}G'_{n+2} + (G_{m+1} + G_{m+2})(G'_{n+1} + G'_{n+2})$$

$$= G_{m+1}G'_{n+1} + G_{m+1}G'_{n+2} + G_{m+2}G'_{n+1} + 2G_{m+2}G'_{n+2}.$$

Combining these two results, it is clear that

$$2(G_{m+2}G'_{n+2} + G_{m+3}G'_{n+3}) + (G_{m+1}G'_{n+2} + G_{m+2}G'_{n+1})$$
$$= 2G_{m+1}G'_{n+1} + 3G_{m+1}G'_{n+2} + 3G_{m+2}G'_{n+1} + 4G_{m+2}G'_{n+2}$$
$$= G_{m+3}G'_{n+4} + G'_{n+3}G_{m+4}.$$

So another sum-of-products formula is

$$G_{m+3}G'_{n+4} + G'_{n+3}G_{m+4} = 2(G_{m+2}G'_{n+2} + G_{m+3}G'_{n+3}) + (G_{m+1}G'_{n+2} + G_{m+2}G'_{n+1}).$$

By allowing G and G' to be the same sequence, or allowing m and n to be equal, or doing both, three special cases of the above formula are obtained:

$$G_{m+3}G_{n+4} + G_{n+3}G_{m+4} = 2(G_{m+2}G_{n+2} + G_{m+3}G_{n+3}) +$$
$$(G_{m+1}G_{n+2} + G_{m+2}G_{n+1}),$$

$$G_{m+3}G'_{m+4} + G'_{m+3}G_{m+4} = 2(G_{m+2}G'_{m+2} + G_{m+3}G'_{m+3}) +$$
$$(G_{m+1}G'_{m+2} + G_{m+2}G'_{m+1}), \text{ and}$$

$$G_{m+3}G_{m+4} = (G_{m+2}^2 + G_{m+3}^2) + (G_{m+1}G_{m+2}).$$

(In the last of these, what is actually obtained is double on both sides.) Alternatively, this could be written as a *difference* of sums of products:

$$G_{m+3}G'_{n+4} + G'_{n+3}G_{m+4} - (G_{m+1}G'_{n+2} + G_{m+2}G'_{n+1}) =$$
$$2(G_{m+2}G'_{n+2} + G_{m+3}G'_{n+3}),$$

$$G_{m+3}G_{n+4} + G_{n+3}G_{m+4} - (G_{m+1}G_{n+2} + G_{m+2}G_{n+1}) =$$
$$2(G_{m+2}G_{n+2} + G_{m+3}G_{n+3}),$$

$$G_{m+3}G'_{m+4} + G'_{m+3}G_{m+4} - (G_{m+1}G'_{m+2} + G_{m+2}G'_{m+1}) =$$
$$2(G_{m+2}G'_{m+2} + G_{m+3}G'_{m+3}),$$

$$G_{m+3}G_{m+4} - G_{m+2}G_{m+1} = G_{m+2}^2 + G_{m+3}^2.$$

Of course, the last of these is in fact a difference-of-products formula, but quite different from the ones in Chapter 16. In those, the expressions of the form

$$A_p A'_q - A_r A'_s$$

all had $p + q = r + s$, which is not true here.

Yet another set of sum-of-products formulas can be obtained by looking at the term-by-term product results of Chapter 7. For if it is always true that

$$G_{m+3}G'_{n+3} = 2G_{m+2}G'_{n+2} + 2G_{m+1}G'_{n+1} - G_m G'_n,$$

the equation can be rearranged to give

$$G_{m+3}G'_{n+3} + G_m G'_n = 2(G_{m+2}G'_{n+2} + G_{m+1}G'_{n+1}).$$

A more symmetric form would arise if n is replaced by $n - 3$, giving

$$G_{m+3}G'_n + G_m G'_{n-3} = 2(G_{m+2}G'_{n-1} + G_{m+1}G'_{n-2}).$$

And of course, both can be the same sequence, or one or both can be the Fibonacci or Lucas sequence, or the geometric progression of powers of τ or τ'. The resulting forms will be collected with the others in Chapter 21.

In fact, the last of these formulas is really a special case of a more general formula, which will not be derived in general here because the derivation of the general formula is much messier than this one; one does not have a simple recursion formula to start with. It will simply be stated; the reader may want to check it out if he would like some illustration:

$$G_{m+(2k-1)}G'_n + G_m G'_{n-(2k-1)} = F_{2k-1}(G_{m+2}G'_{n-1} + G_{m+1}G'_{n-2}).$$

Note that the number $2k - 1$ appears, which is always odd; a similar formula cannot be obtained for even spacings between the products:

$$G_{m+2k}G'_n + G_m G'_{n-2k}.$$

Chapter 21: Sum-of-products and difference-of-products formulas, collected.

In this chapter, all the sum-of-products and difference-of-products formulas of the preceding chapters are collected for quick reference. In addition, the subscripts are regularized. Because some formulas were derived through different paths than others, it will be noted that the particular subscript choices in different formulas varied; the forms printed in this chapter will follow a more consistent choice of subscripts.

For all of the difference-of-products formulas, A and A' represent arbitrary recurrent sequences, A'' represents the prototypic co-recurrent sequence of A and A', k_1 and k_2 represent the recursion formula coefficients shared by A and A', G and G' represent arbitrary generalized Fibonacci sequences, and F and L represent the Fibonacci and Lucas sequences. Some additional formulas are provided for completeness that were not given in the earlier chapters; it should be noted that all geometric progressions with **common** ratios τ and τ' are also generalized Fibonacci sequences, and so many specific cases where G_m or G'_m (or both) are F_m, L_m, τ^m, or τ'^m are provided as well.

$$A_m A'_n - A_n A'_m = (-k_2)^m A''_{n-m-1}(A_0 A'_1 - A_1 A'_0)$$

$$A_m A'_{m+i} - A_{m+i} A'_m = (-k_2)^m A''_{i-1}(A_0 A'_1 - A_1 A'_0)$$

$$A_m A_{n+i} - A_n A_{m+i} = (-k_2)^m A''_{n-m-1}(A_0 A_{i+1} - A_1 A_i)$$

$$A_m A_{m+n+i} - A_{m+n} A_{m+i} = (-k_2)^m A''_{n-1}(A_0 A_{i+1} - A_1 A_i)$$

$$G_m G'_n - G_n G'_m = (-1)^m F_{n-m}(G_0 G'_1 - G_1 G'_0)$$

$$G_m G'_{m+i} - G_{m+i} G'_m = (-1)^m F_i(G_0 G'_1 - G_1 G'_0)$$

$$G_m G_{n+i} - G_n G_{m+i} = (-1)^m F_{n-m}(G_0 G_{i+1} - G_1 G_i)$$

$$G_m G_{m+n+i} - G_{m+n} G_{m+i} = (-1)^m F_n(G_0 G_{i+1} - G_1 G_i)$$

$$G_{m+2} G_{m+3} - G_{m+1} G_m = G_{m+1}^2 + G_{m+2}^2$$

$$F_m G_n - F_n G_m = (-1)^{m+1} G_0 F_{n-m}$$

$$F_m G_{m+i} - F_{m+i} G_m = (-1)^{m+1} G_0 F_i$$

$$L_m G_n - L_n G_m = (-1)^m F_{n-m}(2G_1 - G_0)$$

$$L_m G_{m+i} - L_{m+i} G_m = (-1)^m F_i(2G_1 - G_0)$$

$$\tau^m G_n - \tau^n G_m = (-1)^m F_{n-m}(G_1 - \tau G_0)$$

$$\tau^m G_{m+i} - \tau^{m+i} G_m = (-1)^m F_{i-1}(G_1 - \tau G_0)$$

$$\tau'^m G_n - \tau'^n G_m = (-1)^m F_{n-m}(G_1 - \tau' G_0)$$

$$\tau'^{m}G_{m+i} - \tau'^{m+i}G_m = (-1)^m F_i(G_1 - \tau'G_0)$$

$$F_m F_{n+i} - F_n F_{m+i} = (-1)^{m-1} F_i F_{n-m}$$

$$F_m F_{m+n+i} - F_{m+n} F_{m+i} = (-1)^{m-1} F_i F_n$$

$$F_{m+2} F_{m+3} - F_{m+1} F_m = F_{m+1}^2 + F_{m+2}^2$$

$$F_m L_n - F_n L_m = 2(-1)^{m-1} F_{n-m}$$

$$F_m L_{m+i} - F_{m+i} L_m = 2(-1)^{m-1} F_i$$

$$L_m L_{n+i} - L_n L_{m+i} = (-1)^m F_{n-m}(2L_{i+1} - L_i)$$

$$L_m L_{m+n+i} - L_{m+n} L_{m+i} = (-1)^m F_n(2L_{i+1} - L_i)$$

$$L_{m+2} L_{m+3} - L_{m+1} L_m = L_{m+1}^2 + L_{m+2}^2$$

$$(A_m A'_n - A_n A'_m)/(A_p A'_q - A_q A'_p) = (-k_2)^{m-p}(A''_{n-m-1}/A''_{q-p-1})$$

$$(A_m A'_{m+n} - A_{m+n} A'_m)/(A_p A'_{p+q} - A_{p+q} A'_p) = (-k_2)^{m-p}(A''_{n-1}/A''_{q-1})$$

$$(A_m A_{n+i} - A_n A_{m+i})/(A_p A_{q+i} - A_q A_{p+i}) = (-k_2)^{m-p}(A''_{n-m-1}/A''_{q-p-1})$$

$$(A_m A_{m+n+i} - A_{m+n} A_{m+i})/(A_p A_{p+q+i} - A_{p+q} A_{p+i}) = (-k_2)^{m-p}(A''_{n-1}/A''_{q-1})$$

$$(G_m G'_n - G_n G'_m)/(G_p G'_q - G_q G'_p) = (-1)^{m-p}(F_{n-m}/F_{q-p})$$

$$(G_m G'_{m+n} - G_{m+n} G'_m)/(G_p G'_{p+q} - G_{p+q} G'_p) = (-1)^{m-p}(F_n/F_q)$$

$$(G_m G_{n+i} - G_n G_{m+i})/(G_p G_{q+i} - G_q G_{p+i}) = (-1)^{m-p}(F_{n-m}/F_{q-p})$$

$$(G_m G_{m+n+i} - G_{m+n} G_{m+i})/(G_p G_{p+q+i} - G_{p+q} G_{p+i}) = (-1)^{m-p}(F_{n-1}/F_{q-1})$$

$$(F_m G_n - F_n G_m)/(F_p G_q - F_q G_p) = (-1)^{m-p}(F_{n-m}/F_{q-p})$$

$$(F_m G_{m+n} - F_{m+n} G_m)/(F_p G_{p+q} - F_{p+q} G_p) = (-1)^{m-p}(F_n/F_q)$$

$$(L_m G_n - L_n G_m)/(L_p G_q - L_q G_p) = (-1)^{m-p}(F_{n-m}/F_{q-p})$$

$$(L_m G_{m+n} - L_{m+n} G_m)/(L_p G_{p+q} - L_{p+q} G_p) = (-1)^{m-p}(F_{n-1}/F_{q-1})$$

$$(\tau^m G_n - \tau^n G_m)/(\tau^p G_q - \tau^q G_p) = (-1)^{m-p}(F_{n-m}/F_{q-p})$$

$$(\tau^m G_{m+n} - \tau^{m+n} G_m)/(\tau^p G_{p+q} - \tau^{p+q} G_p) = (-1)^{m-p}(F_n/F_q)$$

$$(\tau'^m G_n - \tau'^n G_m)/(\tau'^p G_q - \tau'^q G_p) = (-1)^{m-p}(F_{n-m}/F_{q-p})$$

$$(\tau'^m G_{m+n} - \tau'^{m+n} G_m)/(\tau'^p G_{p+q} - \tau'^{p+q} G_p) = (-1)^{m-p}(F_{n-1}/F_{q-1})$$

$$(F_m F_{n+i} - F_n F_{m+i})/(F_p F_{q+i} - F_q F_{p+i}) = (-1)^{m-p}(F_{n-m}/F_{q-p})$$

$$(F_m F_{m+n+i} - F_{m+n} F_{m+i})/(F_p F_{p+q+i} - F_{p+q} F_{p+i}) = (-1)^{m-p}(F_n/F_q)$$

$$(L_m L_{n+i} - L_n L_{m+i})/(L_p L_{q+i} - L_q L_{p+i}) = (-1)^{m-p}(F_{n-m}/F_{q-p})$$

$$(L_m L_{m+n+i} - L_{m+n} L_{m+i})/(L_p L_{p+q+i} - L_{p+q} L_{p+i}) = (-1)^{m-p}(F_n/F_q)$$

$$G_m G_{m+2n} - G_{m+n}^2 = (-1)^m F_n^2 (G_0 G_2 - G_1^2)$$

$$G_m G_{m+2n+1} - G_{m+n} G_{m+n+1} = (-1)^m F_n F_{n+1} (G_0 G_2 - G_1^2)$$

$$F_m F_{m+2n} - F_{m+n}^2 = (-1)^{m-1} F_n^2$$

$$F_m F_{m+2n+1} - F_{m+n} F_{m+n+1} = (-1)^{m-1} F_n F_{n+1}$$

$$L_m L_{m+2n} - L_{m+n}^2 = 5(-1)^m F_n^2$$

$$L_m L_{m+2n+1} - L_{m+n} L_{m+n+1} = 5(-1)^m L_n L_{n+1}$$

For the sum-of-products formulas, A' and A'' are differently defined (as specified in Chapter 20) as co-recurrent sequences with A such that $A'_0 = 1$, $A'_1 = 0$, $A''_0 = 0$, and $A''_1 = 1$. In common with the difference-of-products formulas, k_1 and k_2 represent the recursion formula coefficients shared by A, A', and A'', G and G' represent arbitrary generalized Fibonacci sequences, and F and L represent the **Fibonacci** and Lucas sequences.

$$A'_n A_0 + A''_n A_1 = A_n$$

$$A'_n A_m + A''_n A_{m+1} = A_{m+n}$$

$$G_{m+3} G'_{n+3} + G_m G'_n = 2(G_{m+2} G'_{n+2} + G_{m+1} G'_{n+1})$$

$$G_{m+3} G'_n + G_m G'_{n-3} = 2(G_{m+2} G'_{n-1} + G_{m+1} G'_{n-2})$$

$$G_{m+(2k-1)} G'_n + G_m G'_{n-(2k-1)} = F_{2k-1}(G_{m+2} G'_{n-1} + G_{m+1} G'_{n-2})$$

$$G_{m+3} G'_{n+4} + G'_{n+3} G_{m+4} = 2(G_{m+2} G'_{n+2} + G_{m+3} G'_{n+3}) + (G_{m+1} G'_{n+2} + G_{m+2} G'_{n+1})$$

$$G_{m+3} G'_{m+4} + G'_{m+3} G_{m+4} = 2(G_{m+2} G'_{m+2} + G_{m+3} G'_{m+3}) + (G_{m+1} G'_{m+2} + G_{m+2} G'_{m+1})$$

$$G_{m+3} G_{n+3} + G_m G_n = 2(G_{m+2} G_{n+2} + G_{m+1} G_{n+1})$$

$$G_{m+3} G_n + G_m G_{n-3} = 2(G_{m+2} G_{n-1} + G_{m+1} G_{n-2})$$

$$G_{m+(2k-1)} G_n + G_m G_{n-(2k-1)} = F_{2k-1}(G_{m+2} G_{n-1} + G_{m+1} G_{n-2})$$

$$G_{m+3} G_{n+4} + G_{n+3} G_{m+4} = 2(G_{m+2} G_{n+2} + G_{m+3} G_{n+3}) + (G_{m+1} G_{n+2} + G_{m+2} G_{n+1})$$

$$F_{n-1} G_0 + F_n G_1 = G_n$$

$$F_{n-1} G_m + F_n G_{m+1} = G_{m+n}$$

$$F_n G_n + F_{n+1} G_{n+1} = G_{2n+1}$$

$$G_{m+3} F_n + G_m F_{n-3} = 2(G_{m+2} F_{n-1} + G_{m+1} F_{n-2})$$

$$G_{m+3} F_{n+3} + G_m F_n = 2(G_{m+2} F_{n+2} + G_{m+1} F_{n+1})$$

$$G_{m+3} F_{n+4} + F_{n+3} G_{m+4} = (G_{m+1} F_{n+2} + G_{m+2} F_{n+1}) + 2(G_{m+2} F_{n+2} + G_{m+3} F_{n+3})$$

$$G_{m+3} F_{m+4} + F_{m+3} G_{m+4} = (G_{m+1} F_{m+2} + G_{m+2} F_{m+1}) + 2(G_{m+2} F_{m+2} + G_{m+3} F_{m+3})$$

$$G_{m+3} L_{n+3} + G_m L_n = 2(G_{m+2} L_{n+2} + G_{m+1} L_{n+1})$$

$$G_{m+3}L_n + G_mL_{n-3} = 2(G_{m+2}L_{n-1} + G_{m+1}L_{n-2})$$

$$G_{m+3}L_{n+4} + L_{n+3}G_{m+4} = (G_{m+1}L_{n+2} + G_{m+2}L_{n+1}) + 2(G_{m+2}L_{n+2} + G_{m+3}L_{n+3})$$

$$G_{m+3}L_{m+4} + L_{m+3}G_{m+4} = (G_{m+1}L_{m+2} + G_{m+2}L_{m+1}) + 2(G_{m+2}L_{m+2} + G_{m+3}L_{m+3})$$

$$G_{m+3}\tau^{n+3} + G_m\tau^n = 2(G_{m+2}\tau^{n+2} + G_{m+1}\tau^{n+1})$$

$$G_{m+3}\tau^n + G_m\tau^{n-3} = 2(G_{m+2}\tau^{n-1} + G_{m+1}\tau^{n-2})$$

$$G_{m+3}\tau^{n+4} + G_{m+4}\tau^{n+3} = (G_{m+1}\tau^{n+2} + G_{m+2}\tau^{n+1}) +$$
$$2(G_{m+2}\tau^{n+2} + G_{m+3}\tau^{n+3})$$

$$G_{m+3}\tau'^{n+3} + G_m\tau'^n = 2(G_{m+2}\tau'^{n+2} + G_{m+1}\tau'^{n+1})$$

$$G_{m+3}\tau'^n + G_m\tau'^{n-3} = 2(G_{m+2}\tau'^{n-1} + G_{m+1}\tau'^{n-2})$$

$$G_{m+3}\tau'^{m+4} + G_{m+4}\tau'^{m+3} = (G_{m+1}\tau'^{m+2} + G_{m+2}\tau'^{m+1}) +$$
$$2(G_{m+2}\tau'^{m+2} + G_{m+3}\tau'^{m+3})$$

$$F_{n-1}F_m + F_nF_{m+1} = F_{m+n}$$

$$F_n^2 + F_{n+1}^2 = F_{2n+1}.$$

$$F_{m+3}F_{n+3} + F_mF_n = 2(F_{m+2}F_{n+2} + F_{m+1}F_{n+1})$$

$$F_{m+3}F_n + F_mF_{n-3} = 2(F_{m+2}F_{n-1} + F_{m+1}F_{n-2})$$

$$F_{m+3}F_{n+4} + F_{n+3}F_{m+4} = 2(F_{m+2}F_{n+2} + F_{m+3}F_{n+3}) + (F_{m+1}F_{n+2} + F_{m+2}F_{n+1})$$

$$2F_{n-1} + F_n = L_n$$

$$F_{n-1}L_m + F_nL_{m+1} = L_{m+n}$$

$$F_{m+3}L_{n+3} + F_mL_n = 2(F_{m+2}L_{n+2} + F_{m+1}L_{n+1})$$

$$F_{m+3}L_n + F_mL_{n-3} = 2(F_{m+2}L_{n-1} + F_{m+1}L_{n-2})$$

$$F_{n-1} + F_n\tau = \tau^n$$

$$F_{n-1}\tau^m + F_n\tau^{m+1} = \tau^{m+n}$$

$$F_{m+3}\tau^{n+3} + F_m\tau^n = 2(F_{m+2}\tau^{n+2} + F_{m+1}\tau^{n+1})$$

$$F_{m+3}\tau^n + F_m\tau^{n-3} = 2(F_{m+2}\tau^{n-1} + F_{m+1}\tau^{n-2})$$

$$F_{n-1} + F_n\tau' = \tau'^n$$

$$F_{n-1}\tau'^m + F_n\tau'^{m+1} = \tau'^{m+n}$$

Sum-of-products and difference-of-products formulas, collected.

$$F_{m+3}\tau'^{n+3} + F_m\tau'^n = 2(F_{m+2}\tau'^{n+2} + F_{m+1}\tau'^{n+1})$$

$$F_{m+3}\tau'^n + F_m\tau'^{n-3} = 2(F_{m+2}\tau'^{n-1} + F_{m+1}\tau'^{n-2})$$

$$L_{m+3}L_{n+3} + L_mL_n = 2(L_{m+2}L_{n+2} + L_{m+1}L_{n+1})$$

$$L_{m+3}L_n + L_mL_{n-3} = 2(L_{m+2}L_{n-1} + L_{m+1}L_{n-2})$$

$$L_{m+3}L_{n+4} + L_{n+3}L_{m+4} = 2(L_{m+2}L_{n+2} + L_{m+3}L_{n+3}) + (L_{m+1}L_{n+2} + L_{m+2}L_{n+1})$$

$$L_{m+3}\tau^{n+3} + L_m\tau^n = 2(L_{m+2}\tau^{n+2} + L_{m+1}\tau^{n+1})$$

$$L_{m+3}\tau^n + L_m\tau^{n-3} = 2(L_{m+2}\tau^{n-1} + L_{m+1}\tau^{n-2})$$

$$L_{m+3}\tau'^{n+3} + L_m\tau'^n = 2(L_{m+2}\tau'^{n+2} + L_{m+1}\tau'^{n+1})$$

$$L_{m+3}\tau'^n + L_m\tau'^{n-3} = 2(L_{m+2}\tau'^{n-1} + L_{m+1}\tau'^{n-2})$$

It should be noted that the formula $F_{n-1}G_0 + F_nG_1 = G_n$, given above, rearranged as $G_n = F_{n-1}G_0 + F_nG_1$, can be used as another general-term formula for generalized Fibonacci sequences.

Chapter 22: Skip-term recurrent sequences.

Suppose that A_1, A_2, A_3, ... is a 3-term recurrent sequence with the **recursion** formula

$$A_n = k_1 A_{n-1} + k_2 A_{n-2}.$$

Then this chapter will demonstrate that the sequences

$$A_2, A_4, A_6, \text{ ...}$$

$$A_3, A_6, A_9, \text{ ...}$$

$$A_4, A_8, A_{12}, \text{ ...}$$

$$A_5, A_{10}, A_{15}, \text{ ...}$$

obtained from it by taking every second, third, fourth, ... term of the sequence are also 3-term recurrent sequences, though the recursion formulas for these sequences are different.

Now, one easy way to see this is to note that recurrent sequences are also multiple geometric sequences, and vice versa, as was shown in Chapter 4. For if one puts

$$A_i = a_1 r^i + a_2 r'^i,$$

then

$$A_{ni} = a_1 r^{ni} + a_2 r'^{ni}$$

$$= a_1 (r^n)^i + a_2 (r'^n)^i,$$

which is a double geometric sequence with the same coefficients as those in the double geometric sequence expansion of A_i. So it can in turn be rewritten as a recurrent sequence by the **double**-geometric sequence to recurrent sequence formula of chapter 4. But it somehow seems more satisfying to show that these sequences are recurrent sequences directly, by establishing the **recursion** formula coefficients directly. First, if

$$A_n = k_1 A_{n-1} + k_2 A_{n-2},$$

it is also true that

$$A_{n-1} = k_1 A_{n-2} + k_2 A_{n-3},$$

so

$$A_n = k_1 (k_1 A_{n-2} + k_2 A_{n-3}) + k_2 A_{n-2}$$

$$= (k_1^2 + k_2) A_{n-2} + k_1 k_2 A_{n-3},$$

but since A_{n-2} can also be written as

$$A_{n-2} = k_1 A_{n-3} + k_2 A_{n-4},$$

which can be rearranged as

$$k_1 A_{n-3} = A_{n-2} - k_2 A_{n-4},$$

by substitution for A_{n-3},

$$A_n = (k_1^2 + k_2) A_{n-2} + k_2 (A_{n-2} - k_2 A_{n-4})$$

$$= (k_1^2 + 2k_2) A_{n-2} - k_2^2 A_{n-4}.$$

Thus, if one sets

$$k'_1 = k_1^2 + 2k_2, \text{ and}$$

$$k'_2 = -k_2^2,$$

the recursion formula

$$A_n = k'_1 A_{n-2} + k'_2 A_{n-4}$$

is established.

Since it is now clear that the sequence A_2, A_4, A_6, ... is a recurrent sequence, one can call it A'_1, A'_2, A'_3, ..., and by the same reasoning show that A'_2, A'_4, A'_6, ... (which, of course, means A_4, A_8, A_{12}, ...) also form a recurrent sequence, which can be called A'''_1, A'''_2, A'''_3, ... (The reason for using 3 primes rather than 2 is simply to allow for using the notation A''_1, A''_2, A''_3, ... to represent A_3, A_6, A_9, ... as soon as this is shown to be a recurrent sequence.) Of course, the computation of k'''_1 and k'''_2 from k'_1 and k'_2 is by the same formula as the computation of k'_1 and k'_2 from k_1 and k_2:

$$k'''_1 = k'^2_1 + 2k'_2, \text{ and}$$

$$k'''_2 = -k'^2_2,$$

by combining the two formulas, one has

$$k'''_1 = (k_1^2 + 2k_2)^2 + 2(-k_2^2)$$

$$= k_1^4 + 4k_1^2 k_2 + 4k_2^2 - 2k_2^2$$

$$= k_1^4 + 4k_1^2 k_2 + 2k_2^2$$

, and

$$k'''_2 = -(-k_2^2)^2 = -k_2^4.$$

As noted earlier, the sequence A_3, A_6, A_9, ... has not yet been shown to be a recurrent sequence. To show this, first consider the original recursion formula in the form:

$$A_{n-2} = k_1 A_{n-3} + k_2 A_{n-4},$$

which can be rearranged to give

$$k_1 A_{n-3} = A_{n-2} - k_2 A_{n-4}$$

or

$$A_{n-3} = (1/k_1)A_{n-2} - (k_2/k_1)A_{n-4}.$$

Because A' has been shown to be a recurrent sequence, there is also

$$A_{n-2} = k'_1 A_{n-4} + k'_2 A_{n-6},$$

which can be rearranged to give

$$k'_2 A_{n-6} = A_{n-2} - k'_1 A_{n-4}$$

or

$$A_{n-6} = (1/k'_2)A_{n-2} - (k'_1/k'_2)A_{n-4}$$

The two equations

$$A_{n-3} = (1/k_1)A_{n-2} - (k_2/k_1)A_{n-4} \text{ and}$$

$$A_{n-6} = (1/k'_2)A_{n-2} - (k'_1/k'_2)A_{n-4}$$

can be solved simultaneously for A_{n-2} and A_{n-4}. The result is

$$A_{n-2} = [(k'_1/k'_2)A_{n-3} - (k_2/k_1)A_{n-6}]/[(1/k_1)(k'_1/k'_2) - (1/k'_2)(k_2/k_1)]$$

$$= (k_1 k'_1 A_{n-3} - k_2 k'_2 A_{n-6})/(k'_1 - k_2)$$

and

$$A_{n-4} = [(1/k_1)A_{n-6} - (1/k'_2)A_{n-3}]/[(1/k_1)(k'_1/k'_2) - (1/k'_2)(k_2/k_1)]$$

$$= (k_1 A_{n-3} - k'_2 A_{n-6})/(k'_1 - k_2).$$

But we have already established that

$$A_n = k'_1 A_{n-2} + k'_2 A_{n-4},$$

so by substituting for A_{n-2} and A_{n-4} in this equation, the resultant expression for A_n is

$$A_n = k'_1(k_1 k'_1 A_{n-3} - k_2 k'_2 A_{n-6})/(k'_1 - k_2) + k'_2(k_1 A_{n-3} - k'_2 A_{n-6})/(k'_1 - k_2).$$

Substituting for k'_1 and k'_2,

$$k''_1 = (k_1 k'^2_1 + k_1 k'_2)/(k'_1 - k_2)$$

$$= [(k_1^2 + 2k_2)^2 k_1 - k_1 k_2^2]/[(k_1^2 + 2k_2) - k_2]$$

$$= [(k_1^2 + 2k_2)^2 k_1 - k_1 k_2^2]/(k_1^2 + k_2)$$

$$= [(k_1^4 + 4k_1^2 k_2 + 4k_2^2)k_1 - k_1 k_2^2]/(k_1^2 + k_2)$$

$$= [(k_1^5 + 4k_1^3 k_2 + 4k_1 k_2^2) - k_1 k_2^2]/(k_1^2 + k_2)$$

$$= (k_1^5 + 4k_1^3 k_2 + 3k_1 k_2^2)/(k_1^2 + k_2)$$

$$= k_1(k_1^4 + 4k_1^2k_2 + 3k_2^2)/(k_1^2 + k_2)$$

$$= k_1(k_1^2 + 3k_2)(k_1^2 + k_2)/(k_1^2 + k_2)$$

$$= k_1(k_1^2 + 3k_2)$$

and

$$k''_2 = (-k_2 k'_1 k'_2 - k'_2{}^2)/(k'_1 - k_2)$$

$$= [k_2^3(k_1^2 + 2k_2) - k_2^4]/(k_1^2 + k_2)$$

$$= (k_1^2k_2^3 + k_2^4)/(k_1^2 + k_2)$$

$$= (k_1^2 + k_2)k_2^3/(k_1^2 + k_2)$$

$$= k_2^3,$$

the resulting expression,

$$A_n = k''_1 A_{n-3} + k''_2 A_{n-6},$$

shows that when the sequence when $A''_1, A''_2, A''_3, \ldots$ is defined to represent $A_3, A_6, A_9, \ldots$, it has just been proved to be a recurrent sequence.

It should be noted that while the new recurrent sequences created by skipping terms in the original sequence were shown as A_2, $A_4, A_6, \ldots$ or $A_3, A_6, A_9, \ldots$ or $A_4, A_8, A_{12}, \ldots$ etc., the derivations shown here, coupled with the fact that any **shifted** form of a recurrent sequence is itself a co-recurrent sequence with it, should make it clear that $A_p, A_{p + j}, A_{p + 2j}, \ldots$ is always a sequence **co**-recurrent with $A_j, A_{2j}, A_{3j}, \ldots$, and in each case it will be referred to as a **skip-term sequence** with **interval** j.

While the recurrent sequence property has not been demonstrated for any of the skip-term sequences with larger **i**ntervals than 4, it should be clear that it can be shown thus for all of them. Table 18 below shows the k-values in the recursion formulas for intervals up to 4:

Interval	First recursion coefficient	Second recursion coefficient
1	k_1	k_2
2	$k_1^2 + 2k_2$	$-k_2^2$
3	$k_1(k_1^2 + 3k_2)$	k_2^3
4	$k_1^4 + 4k_1^2k_2 + 2k_2^2$	$-k_2^4$

Table 18: Recursion formula coefficients for skip-term sequences.

Interestingly, the sequence k_1, $k_1^2 + 2k_2$, $k_1(k_1^2 + 3k_2)$, $k_1^4 + 4k_1^2k_2 + 2k_2^2$,

... of first recursion coefficients can be shown to be a recurrent sequence, with the relations:

$$k_1(k_1^2 + 3k_2) = k_1(k_1^2 + 2k_2) + k_2(k_1),$$

$$k_1^4 + 4k_1^2k_2 + 2k_2^2 = k_1[k_1(k_1^2 + 3k_2)] + k_2(k_1^2 + 2k_2),$$

etc., so they form a co-recurrent sequence with the original recurrent sequence. While this is an interesting phenomenon, and could be used as a way to expand Table 18 to larger intervals, a *demonstration* of this is beyond the scope of this book. It has, however, been used to expand Table 19. Similarly, the second recursion coefficients simply form a geometric progression with **common** ratio of – k_2, in fact being simply $(-k_2)^{i-1}$, where i is the interval. Again, a demonstration of this is beyond the scope of this book.

For generalized Fibonacci sequences, both k_1 and k_2 are always 1, so the skip-term sequences derived from generalized Fibonacci sequences have recursion coefficients which depend only on the **intervals**, and because they are simply *numbers*, it is convenient to present a more complete table than Table 18.

Interval	First recursion coefficient	Second recursion coefficient
1	1	1
2	3	−1
3	4	1
4	7	−1
5	11	1
6	18	−1
7	29	1
8	47	−1

Table 19: Recursion formula coefficients for skip-term sequences from generalized Fibonacci sequences.

In fact, the first recursion coefficients may be recognized here as the Lucas sequence. So for generalized Fibonacci sequences, the formula can be written:

$$G_n = L_iG_{n-i} + (-1)^{i-1}G_{n-2i}.$$

In this latter form, the equation seems to be well-known; both

Vajda and Dunlap give the formula in a slightly different but equivalent form:

$$G_{n+m} + (-1)^m G_{n-m} = L_m G_n.$$

However, the more general form of this for all recurrent sequences (not just generalized Fibonacci sequences) does not seem to be nearly as well-known.

Chapter 23: Divisibility of terms in the Fibonacci and Lucas sequences.

This chapter differs from most in the book, as it will be mostly concerned with the Fibonacci sequence, and to some extent with the Lucas sequence, rather than more general recurrent sequences (or even just generalized Fibonacci sequences). The results, however, will make use of the previously-obtained property of generalized Fibonacci sequences, that

$$G_n = L_iG_{n-i} + (-1)^{i-1}G_{n-2i}.$$

Letting $n = 2i$ and taking F for G in this equation,

$$F_{2i} = L_iF_i,$$

where the fact that $F_0 = 0$ allows the dropping of the second term in the equation. But since L_i is always an integer, this means that F_{2i} is *always* a *multiple* of F_i. In turn, one can see by the same reasoning that F_{ki} is *always* a *multiple* of F_i. See Table 20 to illustrate this.

i	F_i	F_{2i}	F_{3i}	F_{4i}	F_{5i}
0	0	0	0	0	0
1	1	1	2	3	5
2	1	3	8	21	55
3	2	8	34	144	610
4	3	21	144	987	6765
5	5	55	610	6765	75025
6	8	144	2584	46368	832040
7	13	377	10946	317811	9227465
8	21	987	46368	2178309	102334155
9	34	2584	196418	14930352	1134903170
10	55	6765	832040	102334155	12586269025

Table 20: The Fibonacci sequence and some skip-term derivatives.

(Since this result relies on the fact that $F_0 = 0$, it does not apply to *all* generalized Fibonacci sequences, but only to the Fibonacci sequence. There is a similar rule for the Lucas sequence, but L_{ki} is a multiple of L_i only if $i = 0$ or 1, or k is *odd*. See Table 21. It might also be noted that F_{ki} is a multiple of L_i only if $i = 1$, or k is *even*.)

i	L_i	L_{2i}	L_{3i}	L_{4i}	L_{5i}
0	2	2	2	2	2
1	1	3	4	7	11
2	3	7	18	47	123

i	L_i	L_{2i}	L_{3i}	L_{4i}	L_{5i}
3	4	18	76	322	1364
4	7	47	322	2207	15127
5	11	123	1364	15127	167761
6	18	322	5778	103682	1860498
7	29	843	24476	710647	20633239
8	47	2207	103682	4870847	228826127
9	76	5778	439204	33385282	2537720636
10	123	15127	1860498	228826127	28143753123

Table 21: The Lucas sequence and some skip-term derivatives.

Now, if F_{ki} is always a multiple of F_i, the fact that multiplication does not depend on order (*i. e.*, multiplication is *commutative*) means that F_{ki} is always a multiple of F_k as well. But in turn this means that if neither F_i nor F_k is 1 (which means neither i nor k is 1 or 2) F_{ki} is *never a prime*. (I have seen this demonstrated in other books, but since the books do not introduce the topic of skip-term sequences, the proofs are always very hard to follow.) Thus one can always be sure that

$$F_{mn} = cF_mF_n,$$

where c is an integer, and in fact that

$$F_{jkmn...} = cF_jF_kF_mF_n...,$$

regardless of how many factors $j, k, m, n, ...$ there are. (This can be used to help factor an arbitrary number known to belong to the Fibonacci sequence.)

Another thing that may be noted is a relationship between the Fibonacci and Lucas sequence terms with corresponding subscripts. (It may be useful to consult Table 2 to see the two sequences in parallel columns.) Starting with the Binet formula for the Fibonacci sequence

$$F_n = [(\tau^n - \tau'^n)\sqrt{5}]/5,$$

coupled with the Binet-like formula for the Lucas sequence

$$L_n = \tau^n + \tau'^n,$$

one can compute the quantity

$$5F_n^2 - L_n^2 =$$

$$(\tau^n - \tau'^n)^2 - (\tau^n + \tau'^n)^2$$

$$= -4\tau^n\tau'^n$$

$$= -4(\tau\tau')^n$$

$$= 4(-1)^{n+1}.$$

A consequence is that F_n and L_n are either both even or both odd for any n (in fact they will both be even when n is a multiple of 3 and odd otherwise), and also that they can have *no* other common factor than 2. For if F_n and L_n had a common factor 3, say, they could be written as $3x$ and $3y$, with x and y being some unspecified integers, and

$$5F_n{}^2 - L_n{}^2 = 5(3x)^2 - (3y)^2$$

$$= 45x^2 - 9y^2$$

$$= 9(5x^2 - y^2)$$

which is a multiple of 9, and thus certainly not ± 4. And of course the same argument applies with 3 being replaced by any other number except 2.

The question of divisibility is dealt with at great length in Vajda's book (see the Bibliography), if the reader wishes to go into it more deeply.

Chapter 24: Summing recurrent sequences.

This chapter is the first of a series of chapters dealing with formulas for determining the sum of a set of terms involving recurrent sequences. This chapter specifically will deal with sums where the individual terms in the sum are simply the terms in a recurrent sequence; in other words, computing the **sum**

$$A_1 + A_2 + A_3 + \dots + A_n,$$

or even the sum of the more general set of terms

$$A_{m+1} + A_{m+2} + A_{m+3} + \dots + A_n$$

(where $n > m$), for any recurrent sequence A. Later chapters will be concerned with more complex items to be summed, including products like

$$A_{m+1}A'_{n+1} + A_{m+2}A'_{n+2} + A_{m+3}A'_{n+3} + \dots + A_{m+p}A'_{n+p}.$$

All the formulas can be put into a form which includes pairs of terms that look alike except that the subscripts relate to the lower boundary for one and the upper boundary for the other, with the signs changed. This is to be expected, because if one calculates a sum from $m + 1$ to n and a sum from $n + 1$ to p, and adds the two expressions, it should give the sum from $m + 1$ to p. So this form makes it possible to check the formulas for reasonability.

As stated, this chapter specifically will deal with sums where the individual terms in the sum are simply the terms in a recurrent sequence. The reader may have seen a similar procedure used for summing a geometric progression. It is based on writing the recursion formula in the **zero**-right-hand-side form defined in Chapter 3. Because this book is mainly concerned with 3-term recurrent sequences, it will be shown for this specific case, but it can be seen to apply to any recurrent sequence regardless of the number of terms in the recursion formula.

Let

$$S = A_{m+1} + A_{m+2} + A_{m+3} + \dots + A_n.$$

Then one can write

$$k_1 S = k_1 A_{m+1} + k_1 A_{m+2} + k_1 A_{m+3} + \dots + k_1 A_n \text{ and}$$

$$k_2 S = k_2 A_{m+1} + k_2 A_{m+2} + k_2 A_{m+3} + \dots + k_2 A_n.$$

Line up the terms:

$$S = A_{m+1} + A_{m+2} + A_{m+3} + \dots + A_n$$

$$k_1 S = \quad k_1 A_{m+1} + k_1 A_{m+2} + \dots + k_1 A_{n-1} + k_1 A_n$$

$$k_2 S = \quad\quad\quad k_2 A_{m+1} + \dots + k_2 A_{n-2} + k_2 A_{n-1} + k_2 A_n.$$

If one subtracts the second and third lines from the first, one has the following:

$$(1 - k_1 - k_2)S = A_{m+1} + A_{m+2} - k_1 A_{m+1}$$

$$+ (A_{m+3} - k_1 A_{m+2} - k_2 A_{m+1})$$

$$+ \dots$$

$$+ (A_n - k_1 A_{n-1} - k_2 A_{n-2})$$

$$- k_1 A_n - k_2 A_{n-1} - k_2 A_n.$$

Regardless of how many lines of the form $(A_p - k_1 A_{p-1} - k_2 A_{p-2})$ there are, they are all equal to 0 because of the recursion formula. So the equation reduces to

$$(1 - k_1 - k_2)S = A_{m+1} + A_{m+2} - k_1 A_{m+1} - k_1 A_n - k_2 A_{n-1} - k_2 A_n$$

$$= A_{m+1} - k_1 A_{m+1} + A_{m+2} - k_1 A_n - k_2 A_n - k_2 A_{n-1}$$

$$= (1 - k_1)A_{m+1} + A_{m+2} - (k_1 + k_2)A_n - k_2 A_{n-1}, \text{ or}$$

$$S = [(1 - k_1)A_{m+1} + A_{m+2} - (k_1 + k_2)A_n - k_2 A_{n-1}]/(1 - k_1 - k_2).$$

Note that this formula works for any 3-term recurrent sequence. For generalized Fibonacci sequences, where $k_1 = k_2 = 1$, this formula simplifies to

$$S = (G_{m+2} - 2G_n - G_{n-1})/(-1)$$

$$= 2G_n + G_{n-1} - G_{m+2}$$

$$= G_n + (G_n + G_{n-1}) - G_{m+2}$$

$$= G_n + G_{n+1} - G_{m+2}$$

$$= G_{n+2} - G_{m+2}.$$

And if the series being summed is the Fibonacci sequence, and one is starting with $m = 0$, which is frequently the case, so $F_{m+2} = F_2 = 1$, the sum is

$$S = F_{n+2} - 1.$$

One should note that the formula $S = G_{n+2} - G_{m+2}$ behaves in an expected way, because if one sums from $m + 1$ to n and then from $n + 1$ to p, the combined total is

$$S = (G_{n+2} - G_{m+2}) + (G_{p+2} - G_{n+2}) = (G_{p+2} - G_{m+2}).$$

Demonstrating the corresponding fact for the sum formula for a recurrent sequence that is not a generalized Fibonacci sequence is not as easy; but of course it can be shown if one does enough algebraic manipulation.

Of course, since a skip-term sequence is a recurrent sequence, as shown in Chapter 22, formulas can also be derived easily for sums of the form $S = A_{m+i} + A_{m+2i} + A_{m+3i} + \ldots + A_n$. They will look more complicated because k_1 and k_2 are replaced by the expressions derived in Chapter 22.

Another approach to summing a recurrent sequence can be taken as well. It is related to the fact that every recurrent sequence can be written as a double geometric sequence. Because of that, one can use the geometric progression sum formulas

$$a + ar + ar^2 + \ldots + ar^n = a(r^{n+1} - 1)/(r - 1)$$

and

$$a + ar + ar^2 + \ldots + l = (rl - a)/(r - 1).$$

Either of these two ways of writing the same formula can be applied to each part of the double geometric sequence, and the two combined. In other words, given a recurrent sequence A, if one puts

$$A_i = a_1 r^i + a_2 r'^i$$

with any of the expressions for a_1 among

$$a_1 = (A_1 - A_0 r')/(r - r')$$
$$= (A_2 - A_1 r')/(r - r')r$$
$$= (k_2 A_0 + r A_1)/(r - r')r$$

and any of the expressions for a_2 among

$$a_2 = (A_0 r - A_1)/(r - r')$$
$$= (A_1 r - A_2)/(r - r')r'$$
$$= (k_2 A_0 - r' A_1)/(r' - r)r'$$

where

$$r = [k_1 + \sqrt{(k_1^2 + 4k_2)}]/2 \text{ and}$$
$$r' = [k_1 - \sqrt{(k_1^2 + 4k_2)}]/2,$$

it is possible to write

$$A_{m+1} + A_{m+2} + A_{m+3} + \ldots + A_n =$$
$$(a_1 r^{m+1} + a_2 r'^{m+1}) + (a_1 r^{m+2} + a_2 r'^{m+2}) + (a_1 r^{m+3} + a_2 r'^{m+3}) + \ldots +$$

$$(a_1r^n + a_2r'^n) =$$

$$a_1r^{m+1} + a_1r^{m+2} + a_1r^{m+3} + \dots + a_1r^n +$$

$$a_2r'^{m+1} + a_2r'^{m+2} + a_2r'^{m+3} + \dots + a_2r'^n =$$

$$a_1(r^{n+1} - r^{m+1})/(r-1) + a_2(r'^{n+1} - r'^{m+1})/(r'-1),$$

where the appropriate expressions for a_1, a_2, r, and r' can be substituted. Of course, if the sequence is a generalized Fibonacci sequence, $r = \tau$ and $r' = \tau'$. For example, using the Binet formula for the Fibonacci sequence, namely

$$F_i = [(\tau^i - \tau'^i)\sqrt{5}]/5,$$

substituting $a_1 = \sqrt{5}/5$ and $a_2 = -\sqrt{5}/5$ gives

$$F_{m+1} + F_{m+2} + F_{m+3} + \dots + F_n =$$

$$\sqrt{5}/5(\tau^{n+1} - \tau^{m+1})/(\tau-1) - \sqrt{5}/5(\tau'^{n+1} - \tau'^{m+1})/(\tau'-1),$$

but since $\tau' - 1 = -\tau$ and $\tau - 1 = -\tau'$, this in turn can be written

$$F_{m+1} + F_{m+2} + F_{m+3} + \dots + F_n =$$

$$-\sqrt{5}/5(\tau^{n+1} - \tau^{m+1})/\tau' + \sqrt{5}/5(\tau'^{n+1} - \tau'^{m+1})/\tau,$$

and in turn using $\tau\tau' = -1$, this becomes

$$F_{m+1} + F_{m+2} + F_{m+3} + \dots + F_n =$$

$$\sqrt{5}/5(\tau^{n+1} - \tau^{m+1})\tau - \sqrt{5}/5(\tau'^{n+1} - \tau'^{m+1})\tau'$$

$$= \sqrt{5}/5(\tau^{n+2} - \tau^{m+2}) - \sqrt{5}/5(\tau'^{n+2} - \tau'^{m+2})$$

$$= \sqrt{5}/5(\tau^{n+2} - \tau'^{n+2} - \tau^{m+2} + \tau'^{m+2})$$

$$= F_{n+2} - F_{m+2},$$

in agreement with the formula derived earlier. While this procedure leads to the same result as the procedure which was used in the earlier part of the chapter, it has an advantage in the derivation of some formulas which cannot be derived as easily using the first procedure, so that both procedures are useful in particular cases.

Chapter 25: Summing term-by-term products of generalized Fibonacci sequences.

This chapter is concerned with sums of the form

$$G_{m+1}G'_{n+1} + G_{m+2}G'_{n+2} + G_{m+3}G'_{n+3} + \dots + G_{m+p}G'_{n+p},$$

including cases where $G = G'$, $m = n$, or both. (In the case of both $G = G'$ and $m = n$, the sum takes the form

$$G_{m+1}^2 + G_{m+2}^2 + G_{m+3}^2 + \dots + G_{m+p}^2.$$

The method of this chapter is unable to produce a formula for the **sum** of such products of terms if either or both sequences are recurrent sequences in general. (A procedure for such sums will be derived in Chapter 28. Another procedure would be to convert the term-by-term product into a single recurrent sequence, based on the ideas in Chapter 7.) However, if the sequences are both **generalized** *Fibonacci sequences,* a simple formula can be derived. In Chapter 20, the formula

$$G_{m+3}G'_{n+4} + G'_{n+3}G_{m+4} - (G_{m+1}G'_{n+2} + G_{m+2}G'_{n+1}) =$$

$$2(G_{m+2}G'_{n+2} + G_{m+3}G'_{n+3})$$

was derived. Replacing m by $m - 1$ and n by $n - 1$, the formula becomes

$$G_{m+2}G'_{n+3} + G'_{n+2}G_{m+3} - (G_m G'_{n+1} + G_{m+1}G'_n) =$$

$$2(G_{m+1}G'_{n+1} + G_{m+2}G'_{n+2}),$$

which can be rearranged to give

$$2(G_{m+1}G'_{n+1} + G_{m+2}G'_{n+2}) =$$

$$G_{m+2}G'_{n+3} + G'_{n+2}G_{m+3} - (G_m G'_{n+1} + G_{m+1}G'_n), \text{ or}$$

$$G_{m+1}G'_{n+1} + G_{m+2}G'_{n+2} =$$

$$\tfrac{1}{2}(G_{m+2}G'_{n+3} + G'_{n+2}G_{m+3}) - \tfrac{1}{2}(G_m G'_{n+1} + G_{m+1}G'_n).$$

Suppose one first considers two terms G_{m+1} and G_{m+2}, in one sequence, and two corresponding terms, G'_{n+1} and G'_{n+2}, in the second. Then the sum

$$G_{m+1}G'_{n+1} + G_{m+2}G'_{n+2}$$

has been shown to be able to be written in a form that will be more useful to this procedure, namely

$$\tfrac{1}{2}(G_{m+2}G'_{n+3} + G'_{n+2}G_{m+3}) - \tfrac{1}{2}(G_m G'_{n+1} + G_{m+1}G'_n).$$

Now, suppose that it has been established that

$$G_{m+1}G'_{n+1} + G_{m+2}G'_{n+2} + G_{m+3}G'_{n+3} + \ldots + G_{m+j}G'_{n+j}$$

$$= \tfrac{1}{2}(G_{m+j}G'_{n+j+1} + G'_{n+j}G_{m+j+1}) - \tfrac{1}{2}(G_m G'_{n+1} + G_{m+1}G'_n)$$

for all $j > 1$ up to $p - 1$. Then

$$G_{m+1}G'_{n+1} + G_{m+2}G'_{n+2} + G_{m+3}G'_{n+3} + \ldots + G_{m+p-1}G'_{n+p-1} + G_{m+p}G'_{n+p}$$

$$= \tfrac{1}{2}(G_{m+p-1}G'_{n+p} + G'_{n+p-1}G_{m+p}) + G_{m+p}G'_{n+p} - \tfrac{1}{2}(G_m G'_{n+1} + G_{m+1}G'_n).$$

It is useful to look at the one part of the right-hand side that includes p:

$$\tfrac{1}{2}(G_{m+p-1}G'_{n+p} + G'_{n+p-1}G_{m+p}) + G_{m+p}G'_{n+p}$$

$$= \tfrac{1}{2}(G_{m+p-1}G'_{n+p} + G'_{n+p-1}G_{m+p} + 2G_{m+p}G'_{n+p})$$

$$= \tfrac{1}{2}(G_{m+p-1}G'_{n+p} + G_{m+p}G'_{n+p} + G'_{n+p-1}G_{m+p} + G_{m+p}G'_{n+p})$$

$$= \tfrac{1}{2}[(G_{m+p-1} + G_{m+p})G'_{n+p} + (G'_{n+p-1} + G'_{n+p})G_{m+p}]$$

$$= \tfrac{1}{2}(G_{m+p+1}G'_{n+p} + G'_{n+p+1}G_{m+p})$$

which is exactly what is necessary to verify the formula for $j = p$. So the formula

$$G_{m+1}G'_{n+1} + G_{m+2}G'_{n+2} + G_{m+3}G'_{n+3} + \ldots + G_{m+p}G'_{n+p}$$

$$= \tfrac{1}{2}(G_{m+p+1}G'_{n+p} + G'_{n+p+1}G_{m+p}) - \tfrac{1}{2}(G_m G'_{n+1} + G_{m+1}G'_n), \text{ or}$$

$$G_{m+1}G'_{n+1} + G_{m+2}G'_{n+2} + G_{m+3}G'_{n+3} + \ldots + G_{m+p}G'_{n+p}$$

$$= \tfrac{1}{2}(G_{m+p+1}G'_{n+p} + G'_{n+p+1}G_{m+p} - G_m G'_{n+1} - G_{m+1}G'_n),$$

for all $p > 1$.

A collection of special cases of this formula follows.

$$G_{m+1}G'_{m+1} + G_{m+2}G'_{m+2} + G_{m+3}G'_{m+3} + \ldots + G_{m+p}G'_{m+p}$$

$$= \tfrac{1}{2}(G_{m+p+1}G'_{m+p} + G'_{m+p+1}G_{m+p}) - \tfrac{1}{2}(G_m G'_{m+1} + G_{m+1}G'_m).$$

$$G_{m+1}G'_{m+1} + G_{m+2}G'_{m+2} + G_{m+3}G'_{m+3} + \ldots + G_{m+p}G'_{m+p}$$

$$= \tfrac{1}{2}(G_{m+p+1}G'_{m+p} + G'_{m+p+1}G_{m+p} - G_m G'_{m+1} - G_{m+1}G'_m).$$

$$G_{m+1}G_{n+1} + G_{m+2}G_{n+2} + G_{m+3}G_{n+3} + \ldots + G_{m+p}G_{n+p}$$

$$= \tfrac{1}{2}(G_{m+p+1}G_{n+p} + G_{n+p+1}G_{m+p}) - \tfrac{1}{2}(G_m G_{n+1} + G_{m+1}G_n).$$

$$G_{m+1}G_{n+1} + G_{m+2}G_{n+2} + G_{m+3}G_{n+3} + \ldots + G_{m+p}G_{n+p}$$

$$= \tfrac{1}{2}(G_{m+p+1}G_{n+p} + G_{n+p+1}G_{m+p} - G_m G_{n+1} - G_{m+1}G_n).$$

$$G_{m+1}^2 + G_{m+2}^2 + G_{m+3}^2 + \ldots + G_{m+p}^2 = G_{m+p+1}G_{m+p} - G_m G_{m+1}.$$

$$G_{m+1}G_{m+2} + G_{m+2}G_{m+3} + G_{m+3}G_{m+4} + \ldots + G_{m+p}G_{m+p+1}$$
$$= \tfrac{1}{2}\left(G_{m+p+1}^2 + G_{m+p}G_{m+p+2}\right) - \tfrac{1}{2}\left(G_m G_{m+2} + G_{m+1}^2\right).$$

$$G_{m+1}G_{m+2} + G_{m+2}G_{m+3} + G_{m+3}G_{m+4} + \ldots + G_{m+p}G_{m+p+1}$$
$$= \tfrac{1}{2}\left(G_{m+p+1}^2 + G_{m+p}G_{m+p+2} - G_m G_{m+2} - G_{m+1}^2\right),$$

$$G_{m+1}G_{m+3} + G_{m+2}G_{m+4} + G_{m+3}G_{m+5} + \ldots + G_{m+p}G_{m+p+2}$$
$$= \tfrac{1}{2}\left(G_{m+p+1}G_{m+p+2} + G_{m+p}G_{m+p+3}\right) - \tfrac{1}{2}\left(G_m G_{m+3} + G_{m+1}G_{m+2}\right).$$

$$G_{m+1}G_{m+3} + G_{m+2}G_{m+4} + G_{m+3}G_{m+5} + \ldots + G_{m+p}G_{m+p+2}$$
$$= \tfrac{1}{2}\left(G_{m+p+1}G_{m+p+2} + G_{m+p+3}G_{m+p} - G_m G_{m+3} - G_{m+1}G_{m+2}\right).$$

An infinite number of similar formulas to the last few, of course, can be derived by letting $n = m + 3$, $m + 4$, ..., etc. But it may be more useful to present cases with G being F and L, and specific values of m. Specifically, if G is F and $m = 0$,

$$F_1^2 + F_2^2 + F_3^2 + \ldots + F_n^2 = F_n F_{n+1} - F_0 F_1.$$

But since $F_0 = 0$,

$$F_1^2 + F_2^2 + F_3^2 + \ldots + F_n^2 = F_n F_{n+1}.$$

Similarly, if G is L and $m = 0$,

$$L_1^2 + L_2^2 + L_3^2 + \ldots + L_n^2 = L_n L_{n+1} - L_0 L_1.$$

But since $L_0 = 2$ and $L_1 = 1$,

$$L_1^2 + L_2^2 + L_3^2 + \ldots + L_n^2 = L_n L_{n+1} - 2.$$

Of course, since $L_0 = 2$ so $L_0^2 = 4$,

$$L_0^2 + L_1^2 + L_2^2 + \ldots + L_n^2 = L_n L_{n+1} + 2.$$

Additional formulas include:

$$F_1 F_2 + F_2 F_3 + F_3 F_4 + \ldots + F_n F_{n+1}$$
$$= \tfrac{1}{2}\left(F_{n+1}^2 + F_n F_{n+2} - 1\right),$$

$$F_1 F_3 + F_2 F_4 + F_3 F_5 + \ldots + F_n F_{n+2}$$

$$= \tfrac{1}{2}(F_{n+1}F_{n+2} + F_nF_{n+3} - 1).$$

$$L_1L_2 + L_2L_3 + L_3L_4 + \ldots + L_nL_{n+1}$$

$$= \tfrac{1}{2}(L_{n+1}^{2} + L_nL_{n+2} - 7),$$

$$L_1L_3 + L_2L_4 + L_3L_5 + \ldots + L_nL_{n+2}$$

$$= \tfrac{1}{2}(L_{n+1}L_{n+2} + L_nL_{n+3} - 11).$$

Chapter 26:   Summing term-by-term products of geometric progressions with recurrent sequences.

This chapter is concerned with sums of the form

$$P_{m+1}A_{n+1} + P_{m+2}A_{n+2} + P_{m+3}A_{n+3} + \dots + P_{m+p}A_{n+p},$$

where P is a geometric progression (*i. e.*, $P_{m+1} = rP_m$) and A is a 3-term recurrent sequence (*i. e.*, $A_{n+2} = k_1A_{n+1} + k_2A_n$). Summing such a term-by-term product is almost as easy as summing the recurrent sequence itself, which was done in Chapter 24. The only difference is that rather than using the relationship $A_{n+2} - k_1A_{n+1} - k_2A_n = 0$, the relationship used is $P_{m+3}A_{n+3} - k_1rP_{m+2}A_{n+2} - k_2r^2P_{m+1}A_{n+1} = 0$. It can be seen that this is so can be seen by writing

$$P_{m+3}A_{n+3} - k_1rP_{m+2}A_{n+2} - k_2r^2P_{m+1}A_{n+1}$$

$$= P_{m+3}A_{n+3} - k_1(rP_{m+2})A_{n+2} - k_2(r^2P_{m+1})A_{n+1}$$

$$= P_{m+3}A_{n+3} - k_1P_{m+3}A_{n+2} - k_2P_{m+3}A_{n+1}$$

$$= P_{m+3}(A_{n+3} - k_1A_{n+2} - k_2A_{n+1})$$

$$= 0.$$

It will be convenient to write $X_j = P_{m+j}A_{n+j}$ for all j, and it has just been shown that the X's in fact can be considered a recurrent sequence, with the **recursion** formula

$$X_n = k_1rX_{n-1} + k_2r^2X_{n-2},$$

so that the process of Chapter 24 for summing the sequence produces the three lines

$$S = X_1 + X_2 + X_3 + \dots + X_p,$$

$$k_1rS = k_1rX_1 + k_1rX_2 + \dots + k_1rX_{p-1} + k_1rX_p, \text{ and}$$

$$k_2r^2S = k_2r^2X_1 + \dots + k_2r^2X_{p-2} + k_2r^2X_{p-1} + k_2r^2X_p.$$

The same procedure as in Chapter 24 leads to the result

$$S = [(1 - k_1r)X_1 + X_2 - (k_1r + k_2r^2)X_p - k_2r^2X_{p-1}]/(1 - k_1r - k_2r^2)$$

$$= [(1 - k_1r)P_{m+1}A_{n+1} + P_{m+2}A_{n+2}$$
$$- (k_1r + k_2r^2)P_{m+p}A_{n+p} - k_2r^2 P_{m+p-1}A_{n+p-1}]/(1 - k_1r - k_2r^2).$$

Noting that P was defined as a geometric progression (*i. e.*, $P_{m+1} = rP_m$, so $P_{m+i} = r^{i-1}$) and A was defined as a 3-term recurrent sequence (*i. e.*, $A_{n+2} = k_1A_{n+1} + k_2A_n$), this can be rewritten:

$$A_{n+1} + rA_{n+2} + r^2A_{n+3} + \dots + r^{p-1}A_{n+p} =$$

$$[(1 - k_1r)A_{n+1} + rA_{n+2} - (k_1r + k_2r^2)r^{p-1}A_{n+p} - k_2r^pA_{n+p-1}]/(1 - k_1r - k_2r^2).$$

Of course, if $P_{m+i} = ar^{i-1}$, then the factor a would multiply the whole equation. This formula is relatively complex, but the derivation is straightforward, as seen above. Of course, if $r = 1$, the result is the same as the result of the calculation in Chapter 24, allowing for the difference in the notation. Another useful case is $r = -1$, giving the sum of such as

$$A_{n+1} - A_{n+2} + A_{n+3} +- \ldots + (-1)^{p+1}A_{n+p}.$$

From the basic formula this is

$$A_{n+1} - A_{n+2} + A_{n+3} +- \ldots + (-1)^{p+1}A_{n+p} = \{(1 + k_1)A_{n+1} - A_{n+2} +$$
$$(-1)^{p+1}[(k_1 - k_2)A_{n+p} - k_2 A_{n+p-1}]\}/(1 + k_1 - k_2).$$

Reviewing the formulas devised in this chapter, and the special cases that arise by letting A be G, F, or L:

$$A_{n+1} + rA_{n+2} + r^2 A_{n+3} + \ldots + r^{p-1}A_{n+p} =$$
$$[(1 - k_1 r)A_{n+1} + rA_{n+2} - (k_1 r + k_2 r^2)r^{p-1}A_{n+p} - k_2 r^p A_{n+p-1}]/(1 - k_1 r - k_2 r^2)$$

$$G_{n+1} + rG_{n+2} + r^2 G_{n+3} + \ldots + r^{p-1}G_{n+p} =$$
$$[(1 - r)G_{n+1} + rG_{n+2} - (r + r^2)r^{p-1}G_{n+p} - r^p G_{n+p-1}]/(1 - r - r^2)$$

$$F_{n+1} + rF_{n+2} + r^2 F_{n+3} + \ldots + r^{p-1}F_{n+p} =$$
$$[(1 - r)F_{n+1} + rF_{n+2} - (r + r^2)r^{p-1}F_{n+p} - r^p F_{n+p-1}]/(1 - r - r^2)$$

$$L_{n+1} + rL_{n+2} + r^2 L_{n+3} + \ldots + r^{p-1}L_{n+p} =$$
$$[(1 - r)L_{n+1} + rL_{n+2} - (r + r^2)r^{p-1}L_{n+p} - r^p L_{n+p-1}]/(1 - r - r^2)$$

$$A_{n+1} - A_{n+2} + A_{n+3} +- \ldots + (-1)^{p+1}A_{n+p} = \{(1 + k_1)A_{n+1} - A_{n+2} +$$
$$(-1)^{p+1}[(k_1 - k_2)A_{n+p} - k_2 A_{n+p-1}]\}/(1 + k_1 - k_2)$$

$$G_{n+1} - G_{n+2} + G_{n+3} +- \ldots + (-1)^{p+1}G_{n+p} = 2G_{n+1} - G_{n+2} + (-1)^p G_{n+p-1}$$
$$F_{n+1} - F_{n+2} + F_{n+3} +- \ldots + (-1)^{p+1}F_{n+p} = 2F_{n+1} - F_{n+2} + (-1)^p F_{n+p-1}$$
$$L_{n+1} - L_{n+2} + L_{n+3} +- \ldots + (-1)^{p+1}L_{n+p} = 2L_{n+1} - L_{n+2} + (-1)^p L_{n+p-1}$$

Chapter 27: An interesting special case of the geometric progression/generalized Fibonacci sequence product.

At the end of Chapter 24, a procedure for summing recurrent sequences by expanding them as double geometric sequences was introduced. This is particularly useful for sums of the type treated in Chapter 26, involving the term-by-term products of a geometric progression and a generalized Fibonacci sequence, because when the geometric progression has a *common* ratio small enough, the sequence can be summed to infinity by using the infinite geometric progression formula $S = a/(1 - r)$. In particular, any generalized Fibonacci sequence has the form of a double geometric sequence with common ratios τ and τ', and in turn as one goes to later and later terms of the sequence, the progression with common ratio τ dominates (as mentioned in Chapter 4), so that it resembles a geometric progression with common ratio τ closely. Therefore any generalized Fibonacci sequence G multiplied by a geometric progression whose *common* ratio is $1/q$, where $q > \tau$, converges when summed to infinity. An example of some formulas of this type that will not be derived here, but can be demonstrated if it is desired, is

$$F_1/2 + F_2/4 + F_3/8 + \ldots + F_n/2^n + \ldots = 2.$$

Table 22 shows some of the numeric values of sums of this type.

q	$\text{Sum}(F_i/q^i)$	$\text{Sum}(L_i/q^i)$
2	2	4
3	3/5	1
4	4/11	6/11
5	5/19	7/19
6	6/29	8/29
7	7/41	9/41
8	8/55	2/11
9	9/71	11/71
-2	-2/5	0
-3	-3/11	-1/11
-4	-4/19	-2/19

Table 22: Some infinite sums of Fibonacci/Lucas sequences divided by powers.

It can be seen (although no attempt will be made to prove it here) that the sum in the Fibonacci column is $S_F = q/(q^2 - q - 1)$ and the sum in the Lucas column is $S_L = (q + 2)/(q^2 - q - 1)$. (In fact, if q is something other than an integer, as long as $|q| > \tau$, these formulas will hold.) And in turn, the expression $q^2 - q - 1$ is the expression found in the Fibonacci quadratic equation (Chapter 5), with q for r. Not surprisingly, the closer q comes to τ, the nearer the denominator $q^2 - q - 1$ comes to 0, and thus the larger the sum in question becomes. (In fact, as q becomes very close to τ, the sums in question resemble the divergent sum $1 + 1 + 1 + ...$, and as q becomes very close to $-\tau$, the sums in question resemble the oscillatory sum $1 - 1 + 1 -+ ...$, in keeping with this divergence.)

Because every generalized Fibonacci sequence can be expressed as a combination of the Fibonacci and Lucas sequences, as will be shown in Chapter 29, Table 22 (or the formula just given) can be used to give a numeric value for the sum to infinity of any form (G_i/q^i). Since the details depend on the results of Chapter 29, this will not be gone into here, but combining the expression for $G = fF + lL$ there with the two sum formulas for S_F and S_L above will give an expression for the sum to infinity of any such form.

Chapter 28: Summing term-by-term products of recurrent sequences.

This chapter is concerned with sums of the form

$$A_{m+1}A'_{n+1} + A_{m+2}A'_{n+2} + A_{m+3}A'_{n+3} + \dots + A_{m+p}A'_{n+p},$$

where A and A' are 3-term recurrent sequences (*i. e.*, $A_{m+2} = k_1 A_{m+1} + k_2 A_m$, $A'_{n+2} = k_1 A'_{n+1} + k_2 A'_n$).

Unlike Chapters 24 through 26, a final formula for the sum is not going to be exhibited in this chapter. The reason is that the actual formula is so complex that it would be difficult to take in. Rather, the procedure will be explained, and the reader should be able to produce the sum for any specific case.

The first step in the procedure is to use the recurrent sequence to double-geometric sequence formulas of Chapter 4 to convert A to a double geometric sequence. Writing $A_{m+j} = P_j + P'_j$, each of the terms $A_{m+j}A'_{n+j}$ is transformed to a term of the form $(P_j + P'_j) A'_{n+j}$, so that the sum can be written as the total of two sums

$$P_1 A'_{n+1} + P_2 A'_{n+2} + P_3 A'_{n+3} + \dots + P_p A'_{n+p}$$

and

$$P'_1 A'_{n+1} + P'_2 A'_{n+2} + P'_3 A'_{n+3} + \dots + P'_p A'_{n+p},$$

which can each be handled by the method of Chapter 26. It can be seen that the resultant expression for the sum is extremely complex, but the individual steps, as explained in Chapters 4 and 26, can be performed without difficulty.

The complexity of this expression, however, is the reason that the method of Chapter 25 is to be preferred when A and A' are generalized Fibonacci sequences.

Of course, both the method of Chapter 25 and the method described here can be used in the special case where both sequences that are multiplied term by term are identical, which provides a way of summing the squares of terms in any recurrent sequence (and particularly, when the recurrent sequence is a generalized Fibonacci sequence, the method of Chapter 25 can be used, as was noted at the beginning of that chapter).

Since the squares of the terms of a 3-term recurrent sequence do not in turn form a 3-term recurrent sequence (actually, they do form a *4-term* recurrent sequence) the procedure of this chapter needs to be modified to find a way of summing the cubes, fourth powers, etc. of the terms of a 3-term recurrent sequence. In theory it can be done; in practice it is a formidable task except for some specific sequences. (The sum of cubes of the terms of the Fibonacci sequence, for example, can be computed with a formula that can be found in some references.)

In this chapter so far, the assumption has been made that the subscripts of the two sequences increase in the same way. But if A is any recurrent sequence with

$$A_i = k_1 A_{i-1} + k_2 A_{i-2},$$

it is relatively easy to show that the sequence A', where $A'_i = A_{n-i}$ for any given n, is also a recurrent sequence. For

$$A'_i = A_{n-i} = k_1 A_{n-i-1} + k_2 A_{n-i-2}, \text{ so}$$

$$k_2 A_{n-i-2} = A_{n-i} - k_1 A_{n-i-1} = A'_i - k_1 A'_{i+1}, \text{ or}$$

$$A'_{i+2} = A_{n-i-2} = (1/k_2) A'_i - (k_1/k_2) A'_{i+1},$$

a recurrence formula for A'. (Alternatively, if one uses the double geometric sequence form for A, it is even easier, as

$$A'_i = A_{n-i} = a_1 r^{n-i} + a_2 r'^{n-i}$$

$$= (a_1 r^n) r^{-i} + (a_2 r'^n) r'^{-i}$$

$$= (a_1 r^n)(1/r)^i + (a_2 r'^n)(1/r')^i,$$

a proper double geometric sequence with $a_1 r^n$ and $a_2 r'^n$ for the coefficients and $1/r$ and $1/r'$ for the **common** ratios.) Thus any form such as

$$A_{m+1} A'_{n-1} + A_{m+2} A'_{n-2} + A_{m+3} A'_{n-3} + \dots + A_{m+p} A'_{n-p}$$

is just as easy to do as the form

$$A_{m+1} A'_{n+1} + A_{m+2} A'_{n+2} + A_{m+3} A'_{n+3} + \dots + A_{m+p} A'_{n+p},$$

just by transforming the recurrent sequence A' to its reverse as is shown here.

Of course, if the sequences are both generalized Fibonacci sequences, an alternative approach is to convert the product to a sum of **Fibonacci** and Lucas sequence terms, using the method which will be developed in Chapter 34, and then sum the sequences using the formulas of Chapter 24. (This will not be illustrated here, but when the reader has reached Chapter 34, it will be an obvious procedure.)

Chapter 29:   Symmetries of generalized Fibonacci sequences.

The material in this chapter applies only to generalized Fibonacci sequences, rather than recurrent sequences in general, because it depends specifically on the property that $k_1 = k_2$ in these sequences. Since only generalized Fibonacci sequences are being discussed, the coefficients k_1 and k_2 can be taken as 1 and thus ignored. Suppose a generalized Fibonacci sequence contains two consecutive terms G_p and G_{p+1}, and two other terms G_{q-1} and G_q, with $G_p = G_q$ and $G_{p+1} = -G_{q-1}$. Then

$$G_{p+2} = G_p + G_{p+1},$$

while (by rearranging the recursion formula)

$$G_{q-2} = G_q - G_{q-1}$$
$$= G_p - G_{p+1},$$
$$= G_{p+2}.$$

Similarly,

$$G_{p+3} = G_{p+1} + G_{p+2},$$

while

$$G_{q-3} = G_{q-1} - G_{q-2}$$
$$= -G_{p+1} - G_{p+2}$$
$$= -(G_{p+1} + G_{p+2})$$
$$= -G_{p+3}.$$

By continuing the process as many times as necessary, it can be seen that

$$G_{q-n} = (-1)^n G_{p+n}$$

for all n. Alternatively, a similar argument shows that if a generalized Fibonacci sequence contains two consecutive terms G_p and G_{p+1}, and two other terms G_{q-1} and G_q, with $G_p = -G_q$ and $G_{p+1} = G_{q-1}$, then

$$G_{q-n} = (-1)^{n-1} G_{p+n}$$

for all n.

In particular, for any generalized Fibonacci sequence with $G_p = 0$, it can be seen that $G_{p+1} = G_{p-1}$, making this an example of the second of these two formulas (since $0 = -0$), which specifically means that the

Fibonacci sequence has $F_{-n} = (-1)^{n-1}F_n$ for all n. And similarly, for any generalized Fibonacci sequence with $G_p = 2G_{p+1}$, it can be seen that $G_{p-1} = -G_{p+1}$, making this an example of the second of these two formulas, so that the Lucas sequence has $L_{-n} = (-1)^n L_n$ for all n.

There is an interesting consequence of this. It means that *any* generalized Fibonacci sequence can be considered a combination of the Fibonacci and Lucas sequences, multiplied by appropriate coefficients. Suppose the sequence is G, and we wish to determine coefficients f and l such that

$$G_n = fF_n + lL_n$$

for all n. If one notes that $F_0 = 0$ and $L_0 = 2$, it is clear that $l = \frac{1}{2}G_0$. Since it is necessary to have

$$G_1 = fF_1 + \frac{1}{2}G_0 L_1$$

with $F_1 = L_1 = 1$, the equation can be solved for f:

$$G_1 = fF_1 + \frac{1}{2}G_0 L_1$$

$$= f + \frac{1}{2}G_0,$$

$$f = -\frac{1}{2}G_0 + G_1.$$

Putting it all together, this means that any generalized Fibonacci sequence can be expressed as

$$G_n = (-\frac{1}{2}G_0 + G_1)F_n + \frac{1}{2}G_0 L_n$$

for all n. A slightly more symmetric expression is obtained if one uses the fact that $G_1 = G_0 + G_{-1}$ (or in other words $G_0 = G_1 - G_{-1}$) to replace $\frac{1}{2}G_0$ in this expression. The result is

$$G_n = \frac{1}{2}(G_{-1} + G_1)F_n + \frac{1}{2}G_0 L_n.$$

(It may be recalled that in Chapter 9 it was stated that co-recurrent sequences form what is known to mathematicians as a *vector space*; this is a consequence of that fact, since all generalized Fibonacci sequences have the same recursion formula.)

A consequence of the formula $G_n = \frac{1}{2}(G_{-1} + G_1)F_n + \frac{1}{2}G_0 L_n$ is that any generalized Fibonacci sequence for which $G_{-1} = G_1$ (and of necessity $G_0 = 0$) must be the Fibonacci sequence (possibly multiplied by a constant) and any generalized Fibonacci sequence for which $G_{-1} = -G_1$ must be the Lucas sequence (possibly multiplied by a constant).

This fact enables some other symmetries to be discovered. Note that for any generalized Fibonacci sequence, the formula defining it,

$$G_n = G_{n-1} + G_{n-2},$$

can be rearranged to produce

$$G_{n-2} = G_n - G_{n-1},$$

which enables extending the sequence backwards. First, considering the Fibonacci sequence,

$$F_{-1} = F_1 - F_0 = 1 - 0 = 1 = F_1,$$

$$F_{-2} = F_0 - F_{-1} = -F_0 - F_1 = -(F_0 + F_1) = -F_2,$$

$$F_{-3} = F_{-1} - F_{-2} = F_1 + F_2 = F_3$$

$$F_{-4} = F_{-2} - F_{-3} = -F_2 - F_3 = -(F_2 + F_3) = -F_4,$$

$$F_{-5} = F_{-3} - F_{-4} = F_3 + F_4 = F_5,$$

and it can be seen that this process can go on forever, so it can be stated that for all *odd* integers n, $F_{-n} = F_n$ and for all *even* integers n, $F_{-n} = -F_n$. Alternatively, for *any* integer n, $F_{-n} = (-1)^{n-1}F_n$. The Lucas sequence works similarly, but not exactly the same:

$$L_{-1} = L_1 - L_0 = 1 - 2 = -1 = -L_1,$$

$$L_{-2} = L_0 - L_{-1} = L_0 + L_1 = L_2,$$

$$L_{-3} = L_{-1} - L_{-2} = -L_1 - L_2 = -(L_1 + L_2) = -L_3,$$

$$L_{-4} = L_{-2} - L_{-3} = L_2 + L_3 = L_4,$$

$$L_{-5} = L_{-3} - L_{-4} = -L_3 - L_4 = -(L_3 + L_4) = -L_5,$$

and it can be seen that this process can also go on forever. Thus, for all *odd* integers n, $L_{-n} = -L_n$ and for all *even* integers n, $L_{-n} = L_n$. Alternatively, for *any* integer n, $L_{-n} = (-1)^n L_n$.

In a sense, these properties of the Fibonacci sequence (and also the Lucas sequence) contradict one of the purposes that were in mind for writing this book. Since the book was written with the objective of showing that the two sequences were not quite as special as some writers maintain, but are simply special forms of generalized Fibonacci sequence with the specific numbers 0 and 1 (for the Fibonacci sequence) or 2 and 1 put in for G_0 and G_1, the fact that

- **Any generalized Fibonacci sequence G with the property that $|G_{-n}| = |G_n|$ for all n is either the Fibonacci sequence, the Lucas sequence, or a constant multiple of one of these two sequences**

does make these two sequences special in a way. But it is also true that many of the special formulas involving these two sequences found in other books are purely consequences of the fact that they are generalized Fibonacci sequences and all such sequences obey such formulas.

When all these symmetries are considered, some additional

properties come to light. For example, suppose that a sequence is generated from the Fibonacci sequence by $G_n = F_{n-k} + F_{n+k}$. It is clearly a generalized Fibonacci sequence because it is the sum of two shifted Fibonacci sequences. If k is even, then, as was just shown above,

$$G_0 = F_{-k} + F_k = 0,$$

so G is simply a multiple of the Fibonacci sequence. If k is odd, the reasoning is a little more complex, but it can be seen that

$$G_1 = F_{1-k} + F_{1+k} \text{ and}$$

$$G_{-1} = F_{-1-k} + F_{-1+k} = F_{-(1+k)} + F_{-(1-k)},$$

so, since k being odd implies that both $k + 1$ and $k - 1$ are even,

$$G_1 = -G_{-1}$$

and G is simply a multiple of the Lucas sequence. If one carries out the same reasoning for a sequence G defined by $G_n = F_{n-k} - F_{n+k}$, the odd and even cases are versed, with odd k giving a multiple of the Lucas sequence and even k giving a multiple of the Fibonacci sequence. Furthermore, if similar combinations of terms of the Lucas sequence are taken, the same reasoning can be applied, but the positive and negative signs become interchanged. So, putting all these together, it is demonstrated that, for all *odd* k, there are particular constants c (not all the same!) such that

$$F_{n-k} + F_{n+k} = cL_n,$$

$$L_{n-k} + L_{n+k} = cF_n,$$

$$F_{n-k} - F_{n+k} = cF_n, \text{ and}$$

$$L_{n-k} - L_{n+k} = cL_n,$$

while for all *even* k, there are particular constants c such that

$$F_{n-k} + F_{n+k} = cF_n,$$

$$L_{n-k} + L_{n+k} = cL_n,$$

$$F_{n-k} - F_{n+k} = cL_n, \text{ and}$$

$$L_{n-k} - L_{n+k} = cF_n.$$

Specific values of the constants c for particular choices of k will be given in Chapter 30. (In fact, they are all terms of the Fibonacci or Lucas sequence, except when the form is $L_{n-k} \pm L_{n+k} = cF_n$, when c is 5 times a member of the Fibonacci sequence.)

Chapter 30: Fibonacci and Lucas sequence formulas involving pairwise sums and differences.

In the previous chapter, it was shown that, for any generalized Fibonacci sequence G,

$$G_n = \tfrac{1}{2}(G_{-1} + G_1)F_n + \tfrac{1}{2}G_0 L_n.$$

If f and l are defined as

$$f = \tfrac{1}{2}(G_{-1} + G_1) \text{ and}$$

$$l = \tfrac{1}{2}G_0,$$

all generalized Fibonacci sequences can be expressed in the form

$$G_i = fF_i + lL_i.$$

But it was noted in Chapter 3 that any shifted Fibonacci sequence is a generalized Fibonacci sequence (and can be seen that similarly any shifted Lucas sequence is a generalized Fibonacci sequence). Furthermore, since the sum of any two generalized Fibonacci sequences is a generalized Fibonacci sequence, it means that it is always possible to find an expression of the form $fF_i + lL_i$ to represent *any sum* of shifted Fibonacci sequences and/or shifted Lucas sequences, by simply using the formulas just given.

In addition, since one can write $F_{i+m} = f_1 F_i + l_1 L_i$ and $L_{i+n} = f_2 F_i + l_2 L_i$ as well, it is clear that any sum of shifted Fibonacci sequences and/or shifted Lucas sequences can be written in the form $fF_{i+m} + lL_{i+n}$ for any desired value of m and n, not just $fF_i + lL_i$. So, while some books on the Fibonacci and Lucas sequences treat the equation

$$L_n = F_{n-1} + F_{n+1}$$

as an isolated, almost magical, formula, it can be seen to be just one case of a general relationship, which can be listed along with numerous others. This chapter will contain a long list of such formulas, though obviously such a list can never be complete. (Of course, even the defining formulas $F_n = F_{n-1} + F_{n-2}$ and $L_n = L_{n-1} + L_{n-2}$ are of this type, and so they will be included in this list.)

Within the list, sometimes the same relationship will be written in two different (but equivalent) forms. (For example, $F_{n-1} + F_{n+1} = L_n$ and $F_n + F_{n+2} = L_{n+1}$ say the same thing, though for different values of n.) This does not mean that I cannot see the obvious, or that I do not trust *you* to see the obvious. Rather, it is to illustrate two different patterns, so that in the first case it is grouped with other $F_{n-k} + F_{n+k}$ formulas, and in the second with other $F_n + F_{n+k}$ forms.

It should be noted also that the **sum**-of-products and difference-of-products formulas in Chapter 21 can also be used to generate some of the pairwise sum formulas. For example, from a formula like

$$F_{n-1}G_m + F_n G_{m+1} = G_{m+n},$$

one can, by taking a particular n and letting G be either F or L, produce formulas of this type. Thus, in $n = 4$, $F_n = 3$ and $F_{n-1} = 2$, so the equation becomes

$$2F_m + 3F_{m+1} = F_{m+4} \text{ or}$$

$$2L_m + 3L_{m+1} = L_{m+4}.$$

There are an infinite number of such equations, and only a few will be given in this list.

One thing that was pointed out in Chapter 29 should be repeated immediately. Since $F_{-n} = (-1)^{n-1}F_n$ and $L_{-n} = (-1)^n L_n$ for any integer n, it is clear that

$$F_{2k} + F_{-2k} = 0,$$

$$F_{2k-1} - F_{-(2k-1)} = 0,$$

$$L_{2k} - L_{-2k} = 0, \text{ and}$$

$$L_{2k-1} + L_{-(2k-1)} = 0.$$

It was noted in Chapter 29 that any generalized Fibonacci sequence G whose 0-subscript term $G_0 = 0$ must be the Fibonacci sequence (possibly multiplied by a constant), so the sums $F_{n+2k} + F_{n-2k}$, $F_{n+(2k-1)} - F_{n-(2k-1)}$, $L_{n+(2k-1)} + L_{n-(2k-1)}$, and $L_{n+2k} - L_{n-2k}$ can all be written in the form cF_n, while the forms of the same type, but with the signs reversed, can all be written in the form cL_n. Some of these are included in the following list, but obviously, as n can be any integer, there are an infinite number of such identities, so only a sample can be included.

Because this list is rather long, it may help to explain how it is organized so the reader will have an easier time searching for a particular formula. The primary division of the list is into three sections:

1. Identities involving sums or differences of pairs of Fibonacci sequence terms, $c_1F_i \pm c_2F_j$, with i and j chosen as described below.

2. Identities involving sums or differences of Fibonacci and Lucas sequence terms, $c_1F_i \pm c_2L_j$ or $c_1L_i \pm c_2F_j$, with i and j chosen as described below.

3. Identities involving sums or differences of pairs of Lucas sequence terms, $c_1L_i \pm c_2L_j$, with i and j chosen as described

below.

Within each of these groups, the subscripts i and j are either chosen as n and $n - k$ or as $n + k$ and $n - k$, except in the second group, where combinations of F_n and L_n also appear. In that section, the combinations of F_n and L_n are listed first of all; in all three sections the combinations of terms with subscripts n and $n - k$ precede the combinations of terms with subscripts $n + k$ and $n - k$; and within those combinations, they are listed in order of increasing k; sums precede differences. (There is no need to include sets like $F_n + F_{n+k}$; they can be expressed in the form $F_n + F_{n-k}$ with a different choice of n. Similarly, there is no need to include sets like $F_{n+j} + F_{n+k}$; they can also be expressed in the form $F_n + F_{n-k}$ with different choices of k and n. Strictly speaking, there is not any need either to include such sets as the ones involving $F_{n+k} \pm F_{n-k}$ which *are* included, but because of their symmetry it was chosen to include them.)

$$F_n + F_{n-1} = F_{n+1}$$

$$F_n + 2F_{n-1} = L_n$$

$$2F_n + F_{n-1} = F_{n+2}$$

$$3F_n + 2F_{n-1} = F_{n+3}$$

$$5F_n + 3F_{n-1} = F_{n+4}$$

$$F_n + F_{n-2} = L_{n-1}$$

$$F_n - F_{n-2} = F_{n-1}$$

$$F_n + F_{n-3} = 2F_{n-1} = F_{n-2} + L_{n-2}$$

$$F_n - F_{n-3} = 2F_{n-2}$$

$$F_n - 2F_{n-3} = L_{n-3}$$

$$F_n + F_{n-4} = 3F_{n-2}$$

$$F_n - F_{n-4} = L_{n-2}$$

$$F_n + F_{n-5} = F_{n-3} + L_{n-2}$$

$$F_n - F_{n-5} = L_{n-3} + F_{n-2}$$

$$F_n + F_{n-6} = 2L_{n-3}$$

$$F_n - F_{n-6} = 4F_{n-3}$$

$$F_{n+1} + F_{n-1} = L_n$$

$$F_{n+1} - F_{n-1} = F_n$$

$$F_{n+2} + F_{n-2} = 3F_n$$

$$F_{n+2} - F_{n-2} = L_n$$

$$F_{n+3} + F_{n-3} = 2L_n$$

$$F_{n+3} - F_{n-3} = 4F_n$$

$$F_{n+4} + F_{n-4} = 7F_n$$

$$F_{n+4} - F_{n-4} = 3L_n$$

$$F_{n+5} + F_{n-5} = 5L_n$$

$$F_{n+5} - F_{n-5} = 11F_n$$

$$F_{n+6} + F_{n-6} = 18F_n$$

$$F_{n+6} - F_{n-6} = 8L_n$$

$$F_{n+7} + F_{n-7} = 13L_n$$

$$F_{n+7} - F_{n-7} = 29F_n$$

$$F_{n+8} + F_{n-8} = 47F_n$$

$$F_{n+8} - F_{n-8} = 21L_n$$

$$F_{n+9} + F_{n-9} = 34L_n$$

$$F_{n+9} - F_{n-9} = 76F_n$$

$$F_{n+10} + F_{n-10} = 123F_n$$

$$F_{n+10} - F_{n-10} = 55L_n$$

$$L_n + F_n = 2F_{n+1}$$

$$L_n - F_n = 2F_{n-1}$$

$$L_n + 2F_n = F_{n+3}$$

$$L_n - 2F_n = F_{n-3}$$

$$L_n + 3F_n = 2F_{n+2}$$

$$3L_n + 5F_n = 2L_{n+2}$$

$$L_n + L_{n-1} = L_{n+1}$$

$$2L_n + L_{n-1} = L_{n+2}$$

$$3L_n + 2L_{n-1} = L_{n+3}$$

$$5L_n + 3L_{n-1} = L_{n+4}$$

$$L_n + L_{n-2} = 5F_{n-1}$$

$$L_n + L_{n-3} = 2L_{n-1}$$

$$L_n - 2L_{n-3} = 5F_{n-3}$$

$$L_n + L_{n-4} = 3L_{n-2}$$

Fibonacci and Lucas sequence formulas involving pairwise sums and differences.

$$L_n - L_{n-4} = 5F_{n-2}$$

$$L_n + L_{n-6} = 10F_{n-3}$$

$$L_n - L_{n-6} = 4L_{n-3}$$

$$L_{n+1} + L_{n-1} = 5F_n$$

$$L_{n+1} - L_{n-1} = L_n$$

$$L_{n+2} + L_{n-2} = 3L_n$$

$$L_{n+2} - L_{n-2} = 5F_n$$

$$L_{n+3} + L_{n-3} = 10F_n$$

$$L_{n+3} - L_{n-3} = 4L_n$$

$$L_{n+4} + L_{n-4} = 7L_n$$

$$L_{n+4} - L_{n-4} = 15F_n$$

$$L_{n+5} + L_{n-5} = 25F_n$$

$$L_{n+5} - L_{n-5} = 11L_n$$

$$L_{n+6} + L_{n-6} = 18L_n$$

$$L_{n+6} - L_{n-6} = 40F_n$$

$$L_{n+7} + L_{n-7} = 65F_n$$

$$L_{n+7} - L_{n-7} = 29L_n$$

$$L_{n+8} + L_{n-8} = 47L_n$$

$$L_{n+8} - L_{n-8} = 105F_n$$

$$L_{n+9} + L_{n-9} = 170F_n$$

$$L_{n+9} - L_{n-9} = 76L_n$$

$$L_{n+10} + L_{n-10} = 123L_n$$

$$L_{n+10} - L_{n-10} = 275F_n$$

Chapter 31:   Some additional summation formulas.

The formulas of Chapter 30 can be combined with the formulas of Chapter 24 in interesting ways. For example, in Chapter 24, it was shown that

$$G_{m+1} + G_{m+2} + G_{m+3} + ... + G_n = G_{n+2} - G_{m+2}$$

for any generalized Fibonacci sequence. But when the generalized Fibonacci sequence G is the **Fibonacci** or Lucas sequence, depending on the values of m and n, using the formulas of Chapter 30, the difference $G_{n+2} - G_{m+2}$ can be expressed more simply. For example, if m is replaced by $n-1$ and n is replaced by $n+3$, the above formula for $G = F$ becomes

$$F_n + F_{n+1} + F_{n+2} + F_{n+3} = F_{n+5} - F_{n+1}.$$

But it was shown in Chapter 30 that

$$F_n - F_{n-4} = L_{n-2},$$

and, replacing n by $n+5$, this means that $F_{n+5} - F_{n+1} = L_{n+3}$, so

$$F_n + F_{n+1} + F_{n+2} + F_{n+3} = L_{n+3}.$$

By a similar line of reasoning,

$$L_n + L_{n+1} + L_{n+2} + L_{n+3} = 5F_{n+3}.$$

Table 23 shows a few of these summation expressions.

k	$F_n+F_{n+1}+...+F_{n+k}$	$L_n+L_{n+1}+...+L_{n+k}$
1	F_{n+2}	L_{n+2}
2	$2F_{n+2}$	$2L_{n+2}$
3	L_{n+3}	$5F_{n+3}$
5	$4F_{n+4}$	$4L_{n+4}$
7	$3L_{n+5}$	$15F_{n+5}$
9	$11F_{n+6}$	$11L_{n+6}$
11	$8L_{n+7}$	$40F_{n+7}$
13	$29F_{n+8}$	$29L_{n+8}$
15	$21L_{n+9}$	$105F_{n+9}$
17	$76F_{n+10}$	$76L_{n+10}$
19	$55L_{n+11}$	$275F_{n+11}$

Table 23: Expressions for sums of consecutive terms of the Fibonacci and Lucas sequences.

Because not all even values of k give single-term expressions for the entries in Table 23, there are missing rows in this table. When the two entries for a given value of k are identical except for having F in the first column and L in the last (such as the cases $k = 2$ and $k = 5$ in

Table 23 and in fact for all k of the form $4p + 1$, where the coefficient is always F_{2p+1}), then all generalized Fibonacci sequences obey the same formula. This is because one can express any generalized Fibonacci sequence in the form $G = fF + lL$ as was done in Chapter 29:

$$G_n = \tfrac{1}{2}(G_{-1} + G_1)F_n + \tfrac{1}{2}G_0L_n.$$

If one sets $f = \tfrac{1}{2}(G_{-1} + G_1)$ and $l = \tfrac{1}{2}G_0L_n$, then one can put (using $k = 5$ as an example)

$$G_n + G_{n+1} + G_{n+2} + G_{n+3} + G_{n+4} + G_{n+5} =$$

$$f(F_n + F_{n+1} + F_{n+2} + F_{n+3} + F_{n+4} + F_{n+5}) +$$

$$l(L_n + L_{n+1} + L_{n+2} + L_{n+3} + L_{n+4} + L_{n+5})$$

$$= f(4F_{n+4}) + l(4L_{n+4}) = 4(fF_{n+4} + lL_{n+4}) = 4G_{n+4}.$$

In all other cases, the coefficients differ by a factor 5 (being Lucas sequence terms in the first column and 5 times as great in the second) and F and L are interchanged, but one can still find a generalized Fibonacci sequence G' such that $G_n + G_{n+1} + G_{n+2} + G_{n+3} + G_{n+4} + G_{n+5} = G'_{n+k}$, but G' will be a different sequence from G'. For example, suppose G is the sequence with $G_0 = 4$, $G_1 = 3$. Then $G_{-1} = -1$, and the coefficients f and l are

$$f = \tfrac{1}{2}(G_{-1} + G_1) = \tfrac{1}{2}(-1 + 3) = 1 \text{ and } l = \tfrac{1}{2}G_0 = \tfrac{1}{2}(4) = 2,$$

$$i.\,e.,\, G = F + 2L.$$

So, if the case of $k = 3$ is considered,

$$G_n + G_{n+1} + G_{n+2} + G_{n+3} =$$

$$(F_n + F_{n+1} + F_{n+2} + F_{n+3}) + 2(L_n + L_{n+1} + L_{n+2} + L_{n+3}) =$$

$$L_{n+3} + 10F_{n+3},$$

so that if $G'_n = L_n + 10F_n$, it will always be true that

$$G_n + G_{n+1} + G_{n+2} + G_{n+3} = G'_{n+3}.$$

This can be checked by looking at Table 24, where G and G' are tabulated for a few terms.

n	G_n	G'_n
0	4	2
1	3	11
2	7	13
3	10	24
4	17	37
5	27	61
6	44	98

7	71	159
8	115	257
9	186	416
10	301	673
11	487	1089
12	788	1762
13	1275	2851
14	2063	4613

Table 24: A generalized Fibonacci sequence and a related one that gives sums of four consecutive terms of the first.

It can be seen that the sum of any four consecutive entries in the column headed G_n gives the entry in the column headed G'_n immediately to the right of the bottom one of the four (e. g., 10 + 17 + 27 + 44 = 98.)

Chapter 32:   Generalized Fibonacci sequence formulas involving pairwise sums and differences.

In most chapters of this book, formulas involving the **Fibonacci** and Lucas sequences were obtained by deriving the more general formulas that correspond to them for generalized Fibonacci sequences, or even for recurrent sequences in general. The pairwise sum and difference formulas in Chapter 30, however, were derived specifically for the Fibonacci and Lucas sequences, and the reader may wonder whether they can be generalized to generalized Fibonacci sequences or recurrent sequences. It would be hard to produce corresponding formulas for recurrent sequences in general, but by using the expansion formula that expresses any generalized Fibonacci sequence as a sum of multiples of the Fibonacci and Lucas sequences (derived in Chapter 29), many of the formulas of Chapter 30 can be rewritten so as to produce corresponding formulas for generalized Fibonacci sequences. Reviewing that expansion formula, all generalized Fibonacci sequences can be expressed in the form

$$G_i = fF_i + lL_i,$$

where f and l are defined as

$$f = \tfrac{1}{2}(G_{-1} + G_1) \text{ and}$$

$$l = \tfrac{1}{2}G_0,$$

In general, whenever there is a formula that gives

$$c_1F_{n+p} + c_2F_{n+q} = c_3F_{n+r}$$

and

$$c_1L_{n+p} + c_2L_{n+q} = c_3L_{n+r},$$

for the same values of c_1, c_2, c_3, p, q, and r, it is possible to put

$$c_1G_{n+p} + c_2G_{n+q} = c_1(fF_{n+p} + lL_{n+p}) + c_2(fF_{n+q} + lL_{n+q})$$

$$= fc_1F_{n+p} + lc_1L_{n+p} + fc_2F_{n+q} + lc_2L_{n+q}$$

$$= fc_1F_{n+p} + fc_2F_{n+q} + lc_1L_{n+p} + lc_2L_{n+q}$$

$$= f(c_1F_{n+p} + c_2F_{n+q}) + l(c_1L_{n+p} + c_2L_{n+q})$$

$$= fc_3F_{n+r} + lc_3L_{n+r}$$

$$= c_3G_{n+r}.$$

It is true that only a few of the formulas in Chapter 30 take the necessary form, but when they do, this procedure is applicable. For example, suppose that one considers the formulas (from Chapter 30):

$$F_n + F_{n-3} = 2F_{n-1}$$

$$L_n + L_{n-3} = 2L_{n-1}$$

one can see that

$$G_n + G_{n-3} = 2G_{n-1}$$

for any generalized Fibonacci sequence. More often, where one has

$$c_1F_{n+p} + c_2F_{n+q} = c_3F_{n+r},$$

one has

$$c_1L_{n+p} + c_2L_{n+q} = c_3'L_{n+r},$$

where c_3' is different from c_3. And in other cases, one has

$$c_1F_{n+p} + c_2F_{n+q} = c_3L_{n+r},$$

together with

$$c_1L_{n+p} + c_2L_{n+q} = c_3'L_{n+r}.$$

In all these cases, it is possible to derive a formula for

$$c_1G_{n+p} + c_2G_{n+q},$$

but it will involve a different generalized Fibonacci sequence on the right-hand side of the formula, and because the specific generalized Fibonacci sequence that is involved will depend on the specific values of c_3 and c_3', as well as whether F and L are interchanged or not, it will not be possible to give the formulas in a convenient list form as was done in Chapter 30. It is probably better, in this situation, simply to describe the *procedure* for determining the appropriate c_3, G', and r on the right-hand side of the equation

$$c_1G_{n+p} + c_2G_{n+q} = c_3G'_{n+r}$$

for any given c_1, c_2, p, q, and G. For example, let G be the generalized Fibonacci sequence with $G_0 = 3$, $G_1 = 1$, and suppose it is desired to produce a formula of this type for $G_{n+3} + G_{n-3}$. By the use of the generalized Fibonacci sequence definition formula it is possible to calculate $G_{-1} = -2$, so by the expansion formulas of Chapter 29,

$$f = \tfrac{1}{2}(G_{-1} + G_1) = \tfrac{1}{2}(1 - 2) = -\tfrac{1}{2}, \text{ and}$$

$$l = \tfrac{1}{2}G_0 = \tfrac{1}{2}(3) = {}^3/_2.$$

The list in Chapter 30 gives

$$F_{n+3} + F_{n-3} = 2L_n$$

and

$$L_{n+3} + L_{n-3} = 10F_n.$$

So, substituting the appropriate values:

$$G_{n+3} + G_{n-3} = (-\tfrac{1}{2}F_{n+3} + \tfrac{3}{2}L_{n+3}) + (-\tfrac{1}{2}F_{n-3} + \tfrac{3}{2}L_{n-3})$$

$$= -\tfrac{1}{2}F_{n+3} + \tfrac{3}{2}L_{n+3} + -\tfrac{1}{2}F_{n-3} + \tfrac{3}{2}L_{n-3}$$

$$= -\tfrac{1}{2}F_{n+3} + -\tfrac{1}{2}F_{n-3} + \tfrac{3}{2}L_{n+3} + \tfrac{3}{2}L_{n-3}$$

$$= -\tfrac{1}{2}(F_{n+3} + F_{n-3}) + \tfrac{3}{2}(L_{n+3} + L_{n-3})$$

$$= -\tfrac{1}{2}(2L_n) + \tfrac{3}{2}(10F_n)$$

$$= G'_n,$$

where G' is another generalized Fibonacci sequence such that

$$G'_n = 15F_n - L_n,$$

i. e., $G'_0 = -2$, $G'_1 = 14$, $G'_2 = 12$, $G'_3 = 26$, $G'_4 = 38$, $G'_5 = 64$. It would be noted that if another pair of formulas were used rather than the ones for $F_{n+3} + F_{n-3}$ and $L_{n+3} + L_{n-3}$, a different generalized Fibonacci sequence G' would be needed to express the pairwise sum (or difference) of G terms. It will, however, always be possible to express the pairwise sum (or difference) of G terms in terms of *some* generalized Fibonacci sequence G'.

Chapter 33: Symmetric sums and differences of terms in the Fibonacci and Lucas sequences.

This chapter is concerned with **symmetric** sums and differences (i. e., where the subscripts are of the form $n + k$ and $n - k$) of terms in both the **Fibonacci** and Lucas sequences. Table 25 summarizes the formulas for these quantities, as listed in Chapter 30.

k	$F_{n+k} + F_{n-k}$	$F_{n+k} - F_{n-k}$	$L_{n+k} + L_{n-k}$	$L_{n+k} - L_{n-k}$
1	L_n	F_n	$5F_n$	L_n
2	$3F_n$	L_n	$3L_n$	$5F_n$
3	$2L_n$	$4F_n$	$10F_n$	$4L_n$
4	$7F_n$	$3L_n$	$7L_n$	$15F_n$
5	$5L_n$	$11F_n$	$25F_n$	$11L_n$
6	$18F_n$	$8L_n$	$18L_n$	$40F_n$
7	$13L_n$	$29F_n$	$65F_n$	$29L_n$

Table 25: Formulas for symmetric sums and differences of Fibonacci and Lucas sequence terms.

It will be convenient to refer to the sequence (whether F or L) referred to in the header of each column as the **header sequence,** and r L is not the header sequence as the **opposite sequence**. Examining the patterns in Table 25, a number of rules emerge.

1. When k is odd, the expressions are F_kL_n, L_kF_n, $5F_kF_n$, and L_kL_n. Terms in F_n and L_n appear in the columns showing *differences* of the *header* sequences, and in the columns for *sums* of the *opposite* sequences.

2. When k is even, the expressions are L_kF_n, F_kL_n, L_kL_n, and $5F_kF_n$. Terms in F_n and L_n appear in the columns showing *sums* of the *header* sequences, and in the columns for *differences* of the *opposite* sequences.

3. When the header sequence is F, the coefficient of either F_n or L_n is always the k-subscript term of the other sequence. When the header sequence is L, the coefficient of either F_n or L_n is always the k-subscript term of the same sequence, except that if both are F, it is multiplied by 5.

Up to this point, none of these rules has been *proved*, except for values of k from 1 to 7, which have each been worked out individually.

However it is easy to demonstrate that this pattern can be indefinitely extended. Just one example will be given and it should be obvious how to modify the argument for the other cases.

Suppose it has been established that

$$F_{n+k} - F_{n-k} = F_k L_n \text{ and}$$

$$F_{n+(k+1)} + F_{n-(k+1)} = F_{k+1} L_n.$$

Since $F_{n-k} = F_{n-(k+1)} + F_{n-(k+2)}$, it can be asserted that

$$F_{n-(k+2)} = F_{n-k} - F_{n-(k+1)}.$$

But since $F_{n+(k+2)} = F_{n+(k+1)} + F_{n+k}$, it is possible to write

$$F_{n+(k+2)} - F_{n-(k+2)} = [F_{n+(k+1)} + F_{n+k}] - [F_{n-k} - F_{n-(k+1)}]$$

$$= F_{n+(k+1)} + F_{n+k} - F_{n-k} + F_{n-(k+1)}$$

$$= [F_{n+(k+1)} + F_{n-(k+1)}] + (F_{n+k} - F_{n-k})$$

$$= F_{k+1} L_n + F_k L_n$$

$$= (F_{k+1} + F_k)L_n$$

$$= F_{k+2} L_n,$$

exactly what the expected expression for $F_{n+(k+2)} - F_{n-(k+2)}$ ought to be for the rules to hold. Making the appropriate changes, all the following can be established:

When k is odd:

$$F_{n+k} + F_{n-k} = F_k L_n,$$

$$F_{n+k} - F_{n-k} = L_k F_n,$$

$$L_{n+k} + L_{n-k} = 5F_k F_n,$$

$$L_{n+k} - L_{n-k} = L_k L_n.$$

When k is even:

$$F_{n+k} + F_{n-k} = L_k F_n,$$

$$F_{n+k} - F_{n-k} = F_k L_n,$$

$$L_{n+k} + L_{n-k} = L_k L_n,$$

$$L_{n+k} - L_{n-k} = 5F_k F_n.$$

These expressions hold for *all* values of n and k.

Chapter 34:   Product formulas.

At the end of the preceding chapter, eight equations for symmetric sums and differences involving the **Fibonacci** and Lucas sequences were given, four for even values of k and four for odd. When k is odd:

$$F_{n+k} + F_{n-k} = F_k L_n,$$

$$F_{n+k} - F_{n-k} = L_k F_n,$$

$$L_{n+k} + L_{n-k} = 5 F_k F_n,$$

$$L_{n+k} - L_{n-k} = L_k L_n.$$

When k is even:

$$F_{n+k} + F_{n-k} = L_k F_n,$$

$$F_{n+k} - F_{n-k} = F_k L_n,$$

$$L_{n+k} + L_{n-k} = L_k L_n,$$

$$L_{n+k} - L_{n-k} = 5 F_k F_n.$$

It should be noted that these equations can be used in reverse, as well, to express products of terms in the **Fibonacci** and/or Lucas sequences as sums or differences, and when used that way it is better to combine the odd-k and even-k formulas.

$$F_k F_n = [L_{n+k} + (-1)^{k-1} L_{n-k}]/5,$$

$$L_k L_n = L_{n+k} + (-1)^k L_{n-k},$$

$$L_k F_n = F_{n+k} + (-1)^k F_{n-k},$$

$$F_k L_n = F_{n+k} + (-1)^{k-1} F_{n-k}.$$

(Note that, since $F_0 = 0$, the last two of these formulas become $F_n L_n = F_{2n}$ when n and k are equal, an equation which was noted in Chapter 6.) The four formulas, as stated above, appear to be strangely asymmetric because the sign of $(-1)^{k-1}$ or $(-1)^k$ depends on k and not on n. However, this turns out to be because of the use of $n - k$ as a subscript, and if one interchanges k and n, it will be noted that everything is all consistent. One must first note that, for all i, it was already shown that $F_{-i} = (-1)^{i-1} F_i$ and $L_{-i} = (-1)^i L_i$; this applies to $i = n - k$ as well, so one can write

$$F_{k-n} = (-1)^{n-k-1} F_{n-k} \text{ and}$$

$$L_{k-n} = (-1)^{n-k} L_{n-k}.$$

Upon interchanging k and n, and noting that the third and fourth formulas (the ones involving L_kF_n and F_kL_n above) also become interchanged, all the signs remain correct when these facts are all taken into account.

In Chapter 29, it was shown that any generalized Fibonacci sequence G can be expressed in terms of the Fibonacci and Lucas sequences. While it is usually best to use the symmetric form:

$$G_n = \tfrac{1}{2}(G_{-1} + G_1)F_n + \tfrac{1}{2}G_0L_n,$$

for the purposes of this chapter it is better to use the form

$$G_n = (-\tfrac{1}{2}G_0 + G_1)F_n + \tfrac{1}{2}G_0L_n.$$

If a second sequence is similarly expressed:

$$G'_k = (-\tfrac{1}{2}G'_0 + G'_1)F_k + \tfrac{1}{2}G'_0L_k,$$

the product of the two becomes

$$G'_kG_n = [(-\tfrac{1}{2}G'_0 + G'_1)F_k + \tfrac{1}{2}G'_0L_k][(-\tfrac{1}{2}G_0 + G_1)F_n + \tfrac{1}{2}G_0L_n]$$

$$= [(-G'_0 + 2G'_1)F_k + G'_0L_k][(-G_0 + 2G_1)F_n + G_0L_n]/4$$

$$= [(-G'_0 + 2G'_1)(-G_0 + 2G_1)F_kF_n + G'_0(-G_0 + 2G_1)L_kF_n$$

$$+ G_0(-G'_0 + 2G'_1)F_kL_n + G'_0G_0L_kL_n]/4$$

$$= [(G'_0 - 2G'_1)(G_0 - 2G_1)F_kF_n - G'_0(G_0 - 2G_1)L_kF_n$$

$$- G_0(G'_0 - 2G'_1)F_kL_n + G'_0G_0L_kL_n]/4,$$

and, substituting for the products of Fibonacci and Lucas sequence terms,

$$G'_kG_n = \{(G_0 - 2G_1)(G'_0 - 2G'_1)[L_{n+k} + (-1)^{k-1}L_{n-k}]/5$$

$$- G'_0(G_0 - 2G_1)[F_{n+k} + (-1)^kF_{n-k}]$$

$$- G_0(G'_0 - 2G'_1)[F_{n+k} + (-1)^{k-1}F_{n-k}]$$

$$+ G'_0G_0[L_{n+k} + (-1)^kL_{n-k}]\}/4$$

$$= \{-[G'_0(G_0 - 2G_1) + G_0(G'_0 - 2G'_1)]F_{n+k}$$

$$+ [(G_0 - 2G_1)(G'_0 - 2G'_1)/5 + G'_0G_0]L_{n+k}$$

$$+ (-1)^{k-1}[G'_0(G_0 - 2G_1) - G_0(G'_0 - 2G'_1)]F_{n-k}$$

$$+ (-1)^{k-1}[(G_0 - 2G_1)(G'_0 - 2G'_1)/5 - G'_0G_0]L_{n-k}\}/4$$

$$= \{-(G_0G'_0 - 2G_1G'_0 + G_0G'_0 - 2G_0G'_1)F_{n+k}$$

$$+ [(G_0G'_0 - 2G_1G'_0 - 2G_0G'_1 + 4G_1G'_1)/5 + G'_0G_0]L_{n+k}$$

$$+ (-1)^{k-1}[(G_0G'_0 - 2G_1G'_0 - G_0G'_0 + 2G_0G'_1)]F_{n-k}$$
$$+ (-1)^{k-1}[(G_0G'_0 - 2G_1G'_0 - 2G_0G'_1 + 4G_1G'_1)/5 - G'_0G_0]L_{n-k}\}/4$$

$$= \{-(2G_0G'_0 - 2G_1G'_0 - 2G_0G'_1)F_{n+k}$$
$$+ [(6G_0G'_0 - 2G_1G'_0 - 2G_0G'_1 + 4G_1G'_1)/5]L_{n+k}$$
$$+ (-1)^{k-1}[(-2G_1G'_0 + 2G_0G'_1)]F_{n-k}$$
$$+ (-1)^{k-1}[(-4G_0G'_0 - 2G_1G'_0 - 2G_0G'_1 + 4G_1G'_1)/5]L_{n-k}\}/4$$

$$= \{-(G_0G'_0 - G_1G'_0 - G_0G'_1)F_{n+k}$$
$$+ [(3G_0G'_0 - G_1G'_0 - G_0G'_1 + 2G_1G'_1)/5]L_{n+k}$$
$$+ (-1)^{k-1}[(-G_1G'_0 + G_0G'_1)]F_{n-k}$$
$$+ (-1)^{k-1}[(-2G_0G'_0 - G_1G'_0 - G_0G'_1 + 2G_1G'_1)/5]L_{n-k}\}/2$$

$$= \{-(G_0G'_0 - G_1G'_0 - G_0G'_1)F_{n+k}$$
$$+ [(3G_0G'_0 - G_1G'_0 - G_0G'_1 + 2G_1G'_1)/5]L_{n+k}$$
$$+ (-1)^{k-1}[(G_0G'_1 - G_1G'_0)]F_{n-k}$$
$$+ (-1)^{k-1}[(2G_1G'_1 - 2G_0G'_0 - G_1G'_0 - G_0G'_1)/5]L_{n-k}\}/2,$$

which needs to be simplified by looking at the combinations of $G_jG'_k$ that appear as multipliers of the Fibonacci and Lucas terms.

The coefficient $G_0G'_0 - G_1G'_0 - G_0G'_1$ can be written as

$$(G_0 - G_1)G'_0 - G_0G'_1$$
$$= -(G_1 - G_0)G'_0 - G_0G'_1$$
$$= -G_{-1}G'_0 - G_0G'_1$$
$$= -(G_{-1}G'_0 + G_0G'_1).$$

As it already appears with a minus sign multiplying F_{n+k} in the preceding expression, the two signs can be canceled.

The coefficient $3G_0G'_0 - G_1G'_0 - G_0G'_1 + 2G_1G'_1$ can be written as

$$(3G_0 - G_1)G'_0 + (2G_1 - G_0)G'_1$$
$$= [2G_0 + (G_0 - G_1)]G'_0 + [G_1 + (G_1 - G_0)]G'_1$$
$$= [2G_0 - (G_1 - G_0)]G'_0 + [G_1 + (G_1 - G_0)]G'_1$$
$$= (2G_0 - G_{-1})G'_0 + (G_1 + G_{-1})G'_1$$

$$= [G_0 + (G_0 - G_{-1})]G'_0 + (G_1 + G_{-1})G'_1$$

$$= (G_0 + G_{-2})G'_0 + (G_1 + G_{-1})G'_1.$$

The coefficient $G_0 G'_1 - G_1 G'_0$ is already in as simple form as can be used.

The coefficient $2G_1 G'_1 - 2G_0 G'_0 - G_1 G'_0 - G_0 G'_1$ can be written as

$$(2G_1 - G_0)G'_1 - (2G_0 + G_1)G'_0$$

$$= [G_1 + (G_1 - G_0)]G'_1 - [G_0 + (G_0 + G_1)]G'_0$$

$$= (G_1 + G_{-1})G'_1 - (G_0 + G_2)G'_0.$$

Making these substitutions, the equation simplifies to

$$G'_k G_n = \{((G_{-1}G'_0 + G_0 G'_1)F_{n+k}$$

$$+ [(G_0 + G_{-2})G'_0 + (G_1 + G_{-1})G'_1]L_{n+k}/5$$

$$+ (-1)^{k-1}[(G_0 G'_1 - G_1 G'_0)]F_{n-k}$$

$$+ (-1)^{k-1}[(G_1 + G_{-1})G'_1 - (G_0 + G_2)G'_0]L_{n-k}/5\}/2.$$

Note that all the G terms appearing on the right-hand side of this equation do not depend on n or k, so once a particular set of generalized Fibonacci sequences G and G' have been selected, they are constant.

To give an example, suppose the sequence G is given by $G_1 = 3$, $G_2 = 4$, etc., and the sequence G' is given by $G'_1 = 2$, $G'_2 = 5$, etc. In order to use the formulas just given, it is necessary to extend the sequence G back to G_{-2}, and although the sequence G' need only be extended back to G'_0, Table 26 will extend it back as well just for ease in tabulation. (It can be seen that G is simply the Lucas sequence shifted by one place; this is irrelevant to this discussion.)

i	G_i	G'_i
-2	-1	4
-1	2	-1
0	1	3
1	3	2
2	4	5
3	7	7
4	11	12
5	18	19
6	29	31
7	47	50
8	76	81

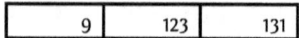

9	123	131

Table 26: Two generalized Fibonacci sequences illustrating the product formula.

The expressions appearing in the coefficients are:

$$G_{-1}G'_0 + G_0G'_1 = (2)(3) + (1)(2) = 8,$$

$$(G_0 + G_{-2})G'_0 + (G_1 + G_{-1})G'_1 = (1 - 1)(3) + (3 + 2)(2) = 10,$$

$$G_0G'_1 - G_1G'_0 = (1)(2) - (3)(3) = -7,$$

$$(G_1 + G_{-1})G'_1 - (G_0 + G_2)G'_0 = (2 + 3)(2) - (1 + 4)(3) = -5.$$

So the formula gives

$$G'_kG_n = [8F_{n+k} + 2L_{n+k} + 7(-1)^kF_{n-k} + (-1)^kL_{n-k}]/2, \text{ or}$$

$$(4F_{n+k} + L_{n+k}) + (-1)^k(7F_{n-k} + L_{n-k})/2.$$

If one checks this formula out by substituting some values, it will be found to be correct. For example, if one puts $k = 5$, $n = 3$:

$$G'_5G_3 = (19)(7) = 133, \text{ and}$$

$$(4F_8 + L_8) - (7F_{-2} + L_{-2})/2 = [(4)(21) + 47] - [(7)(-1) + 3]/2$$

$$= 84 + 47 + 2 = 133.$$

It should also be noted that from this equation one could derive all the difference-of-products formulas given earlier, and from the fact that the two products whose differences are calculated in those difference-of-products formulas have the same (but oppositely-signed) values of $n - k$, the dropping out of certain terms on combining the two products can be seen.

Chapter 35: Congruence cycles in recurrent sequences.

Because congruence (as defined in Chapter 1) is a concept related to integer arithmetic only, for this chapter all references to recurrent sequences will be restricted to sequences with A_0, A_1, k_1, and k_2 restricted to integers. When this is the case, it can easily be seen that all A_p are integers for $p \geq 0$, and the case $p < 0$ will not be considered for the general recurrent sequence A. When A is a generalized Fibonacci sequence G, however, which means $k_1 = k_2 = 1$, it is not even necessary to restrict the consideration to terms G_p with $p \geq 0$, and this restriction will, therefore, not be enforced.

Suppose one considers a recurrent sequence A in which the restriction that A_0, A_1, k_1, and k_2 are integers applies. For any given modulus m, suppose that

$$A_p \equiv A_q \pmod{m} \text{ and}$$

$$A_{p+1} \equiv A_{q+1} \pmod{m},$$

for some p and q. It is then very easy to show that

$$A_{p+2} = k_1 A_p + k_2 A_{p+1} \equiv k_1 A_q + k_2 A_{q+1} = A_{q+2} \pmod{m}.$$

If A is a generalized Fibonacci sequence G, with the conditions that

$$G_p \equiv G_q \pmod{m} \text{ and}$$

$$G_{p+1} \equiv G_{q+1} \pmod{m},$$

for some p and q, it is easily shown not only that

$$G_{p+2} = G_p + G_{p+1} \equiv G_q + G_{q+1} = G_{q+2} \pmod{m},$$

but also that

$$G_{p-1} = G_{p+1} - G_p \equiv G_{q+1} - G_q = G_{q-1} \pmod{m}.$$

[It is pointed out in chapter VII of Vajda's book, using different notation from this one but equivalently, that in recurrent sequences where the second recursion coefficient k_2 is not relatively prime to m, the *backward* extension of this formula, *i. e.*, the one demonstrating that $G_{p-1} \equiv G_{q-1} \pmod{m}$, is not possible for A_{p-1} and A_{q-1} of such sequences, and so the results of this chapter which require that backward extension are restricted to generalized Fibonacci sequences, though in those cases where a modulus is selected that is relatively prime to both recursion coefficients, the same arguments can be applied to recurrent sequences that are not generalized Fibonacci sequences.] Extending this reasoning indefinitely, it is easily seen that

$$A_{p+k} \equiv A_{q+k} \pmod{m},$$

for all k whatever, or, what says the same thing,

$$A_k \equiv A_{k-p+q} \pmod{m},$$

for all k whatever. So the residue classes of the terms of any recurrent sequence A for which two pairs of consecutive terms A_p, A_{p+1}, A_q, and A_{q+1} can be found such that

$$A_p \equiv A_q \pmod{m} \text{ and}$$

$$A_{p+1} \equiv A_{q+1} \pmod{m},$$

under any particular modulus m, form a *cycle*. (When A is not a generalized Fibonacci sequence, it is necessary to exclude the terms A_j where j is less than both p and q, or at least to exclude them until they are tested.) But it is *always* possible to find such pairs. Because there are only m different residue classes under the modulus m, and so there are at most m^2 *pairs* of residue classes of consecutive terms. So any set of consecutive terms

$$A_j, A_{j+1}, A_{j+2}, \dots , A_{j+k},$$

where $k > m^2$, must contain at least two pairs of consecutive terms A_p, A_{p+1}, A_q, and A_{q+1} satisfying the two congruences just given above. In fact, it is usually not necessary to go so far to find such a cycle. Just to give some examples, let $m = 5$, so that $m^2 = 25$. Looking at Table 27, where the Fibonacci, Lucas, and an arbitrarily chosen generalized Fibonacci sequence are tabulated with the residue classes under the modulus 5 are charted, cycles are seen in all 3.

	F_i	$F_i \pmod 5$	L_i	$L_i \pmod 5$	G_i	$G_i \pmod 5$
1	1	1	1	1	4	4
2	1	1	3	3	3	3
3	2	2	4	4	7	2
4	3	3	7	2	10	0
5	5	0	11	1	17	2
6	8	3	18	3	27	2
7	13	3	29	4	44	4
8	21	1	47	2	71	1
9	34	4	76	1	115	0
10	55	0	123	3	186	1
11	89	4	199	4	301	1
12	144	4	322	2	487	2
13	233	3	521	1	788	3

	F_i	$F_i \pmod 5$	L_i	$L_i \pmod 5$	G_i	$G_i \pmod 5$
14	377	2	843	3	1275	0
15	610	0	1364	4	2063	3
16	987	2	2207	2	3338	3
17	1597	2	3571	1	5401	1
18	2584	4	5778	3	8739	4
19	4181	1	9349	4	14140	0
20	6765	0	15127	2	22879	4
21	10946	1	24476	1	37019	4
22	17711	1	39603	3	59898	3
23	28657	2	64079	4	96917	2
24	46368	3	103682	2	156815	0
25	75025	0	167761	1	253732	2

Table 27: Fibonacci and Lucas sequences, and an arbitrary sequence, showing mod-5 values.

It will be seen that the cycle beginning 1, 1, ... for F_i, repeats starting with i = 21; the cycle beginning 1, 3, ... for L_i, repeats starting with i = 5; and the cycle beginning 4, 3, ... for G_i, repeats starting with i = 21 (the same as for F_i). (It might be noted that in the Lucas sequence, there are no zeroes in the mod 5 column, demonstrating that no term of the Lucas sequence is a multiple of 5. Because it has been demonstrated that the sequence 1, 3, 4, 2 repeats indefinitely, this *proves* that no term of the Lucas sequence is a multiple of 5.)

Similar considerations apply to any modulus. Because of the cyclical nature of these congruences, one can also prove by inspection, for example, that the particular G used above satisfies the rule that $G_i + F_{i+1}$ is always a multiple of 5. (In fact, all one would need to show in this case is that $G_1 + F_2 = 4 + 1 = 5$, $G_2 + F_3 = 3 + 2 = 5$, and thus that for all i, $G_i + F_{i+1} = 5F_i$. But in general, such things might be much more difficult without resorting to tables like Table 27 and the cyclical nature of the residues. How, for example, would one show that no Lucas sequence term is a multiple of 5 or of 8?)

Table 28 shows an example of what happens for a recurrent sequence (which is not a generalized Fibonacci sequence) where the modulus is not prime to the k values. The sequence A_n in this table is chosen with k_1 = 3 and k_2 = 4, and, as shown in the table, the modulus is 6. The residues cycle starting with A_1, but A_0 is not in the cycle. This is an example of the general situation, where there may be one or more entries that do not recycle at the beginning of the sequence, but after that point a cycle is obtained. (It would be noted that with the same k values, but setting $A_0 = A_1 = 1$, there would be a complete cycle, with the residues all being 1. So it is not *necessary* that there be a non-cyclic set of entries at the beginning; it is only *possible*.)

n	A_n	A_n (mod 6)
0	1	1
1	2	2
2	10	4
3	38	2
4	154	4
5	614	2
6	2458	4
7	9830	2
8	39322	4
9	157286	2

Table 28: Residue classes (mod 6) for a recurrent sequence with recursion coefficients 3 and 4.

The use of congruence cycles can be used to demonstrate all sorts of interesting facts. For example, since $F_{10} = 55 \equiv 0$ (mod 11) and $F_{11} = 89 \equiv 1$ (mod 11), one can see that $F_{10+n} \equiv F_n$ (mod 11). Or, in other words, $F_{11} \equiv F_1$ (mod 11), $F_{12} \equiv F_2$ (mod 11), etc. Now by the sum formula of Chapter 24,

$$F_1 + F_2 + F_3 + \ldots + F_n = F_{n+2} - 1$$

which means that

$$F_1 + F_2 + F_3 + \ldots + F_{10} = F_{12} - 1 = 143 \equiv 0 \text{ (mod 11)}.$$

If one replaces F_1 by F_{11}, the sum is still going to be 0 (mod 11) because it has already been noted that $F_{11} \equiv F_1$ (mod 11). Similarly, if one replaces F_2 by F_{12}, the sum is still going to be 0 (mod 11) because it has already been noted that $F_{12} \equiv F_2$ (mod 11). And it can be seen that by continuing to replace F_{1+m} by F_{11+m}, where F_{1+m} is the first term of the sum

$$F_{1+m} + F_{2+m} + F_{3+m} + \ldots + F_{10+m},$$

the sum remains 0 (mod 11), which means that the sum of *any 10 consecutive terms* of the Fibonacci sequence is always a multiple of 11! (This same argument implies that if there is a pair of consecutive terms of the Fibonacci sequence F_n and F_{n+1} such that $F_n \equiv 0$ (mod q) and $F_{n+1} \equiv 1$ (mod q), then the sum of any n consecutive terms of the Fibonacci sequence is always a multiple of q. (Sometimes n is rather high for a particular q, as in the case of $q = 5$ where $n = 20$, so one needs to take a lot of terms.)

Chapter 36: Continued fractions.

A **continued fraction** is an expression of the form

$$b_0 + \cfrac{a_1}{b_1 + \cfrac{a_2}{b_2 + \cfrac{a_3}{b_3 + \dots}}}$$

where the expression may continue forever or stop at some particular b_i. Because the format of this expression is rather cumbersome to print, various special ways of writing continued fractions have been devised. It is usual to deal with continued fractions in which all the a's are 1, so it is only necessary to write the b's, and the form

$$b_0 + \cfrac{1}{b_1 + \cfrac{1}{b_2 + \cfrac{1}{b_3 + \dots}}}$$

will be represented as $[b_0; b_1, b_2, b_3, \dots]$ in this book. (Some books do not use the semicolon to distinguish the first number from the others, but I feel it important because this is the only one that is not in a denominator. Other books use other types of brackets; there is no uniformity in this notation.) However, although, strictly speaking, b_0 is not a denominator of anything, it is useful to use the term *"denominators"* to mean the *entire* list of b's in the form above.

Whole books have been written on continued fractions. It is not the purpose of this book to cover the topic in great detail, but some relationships between continued fractions and recurrent sequences are presented in this chapter as matters that may be of interest.

First, it is clear (by expanding the expressions into the standard form) that two numbers represented by $[b_0; b_1, b_2, b_3, \dots]$ and $[0; b_0, b_1, b_2, b_3, \dots]$ are reciprocals, if $b_0 > 0$. So in general, if one has a number expressed, in continued fraction form, as $[b_0; b_1, b_2, b_3, \dots]$, when $b_0 > 0$, its reciprocal is $[0; b_0, b_1, b_2, b_3, \dots]$, and when $b_0 = 0$, its reciprocal is $[b_1; b_2, b_3, \dots]$.

Second, although no attempt will be made to *prove* the following statement, it will become obvious when a procedure is demonstrated to calculate the result: *Any* positive real number can be expressed as a continued fraction, and if it is required to have the fraction in the form $[b_0; b_1, b_2, b_3, \dots, b_n]$, with n no greater than some specified number, then all of b_0, b_1, b_2, b_3, ..., b_{n-1} can be chosen as *integers* (b_0 non-negative, b_1, b_2, b_3, ..., b_{n-1} strictly positive). In general, it will be impossible to force b_n to be an integer, though if the number in question is *rational*, there will be some n that will make b_n integral. (By allowing continued fractions that do not terminate, but run on indefinitely, a form with *all* of the b's integers is possible.)

A brute-force procedure for determining the continued fraction expression for a specified positive real number will now be given. For certain types of numbers, especially rational numbers and quadratic expressions of the form $(a \pm b\sqrt{c})/d$, there are special methods which are preferable, but this method will work for all positive real numbers.

Let x be the number for which a continued fraction expression is desired. Since x has been assumed positive, there is some integer b_0 such that $b_0 \leq x < b_0 + 1$. If $b_0 = x$, simply set $x = [b_0]$ and terminate the process.

Otherwise, put $y_1 = x - b_0$ and $x_1 = 1/y_1$. Since $b_0 \leq x < b_0 + 1$, and we already excluded the case that $b_0 = x$, it is clear that $0 < y_1 < 1$, so $x_1 > 1$. Then the same process can be used to find b_1 from x_1 that was used to find b_0 from x. However, in this case, it is clear that b_1 is at least 1.

Then determine $y_2 = x_1 - b_1$ and $x_2 = 1/y_2$. Continue this process until x_n has been determined, and set $b_n = x_n$. Put the b's into the continued fraction $[b_0; b_1, b_2, b_3, ..., b_n]$, and it will be easily demonstrated that this will have the value of x. The procedure described will produce all the $b_0 ... b_{n-1}$ as integers, $b_0 \geq 0$, $b_1 ... b_{n-1} > 0$, which is what was sought.

It should be noted that since each y value, and specifically y_n, is strictly less than 1, there will never, by this procedure, be a continued fraction generated for which $b_n = 1$. Of course, nothing prevents anyone from *writing down* a continued fraction as $[b_0; b_1, b_2, ..., b_{n-1}, b_n]$, with $b_n = 1$. But whenever a number is found that could be written in that form, it is easily seen that both

$$[b_0; b_1, b_2, ..., b_{n-1}, 1]$$

and

$$[b_0; b_1, b_2, ..., (b_{n-1}+ 1)]$$

represent the same number. And the latter form is what this procedure generates. However, as one will see below, continued fractions with their last denominator 1 do occur as *convergents*, so they can be met with.

Let us use the procedure on the specific number τ, just to show the reason that continued fractions are important to this discussion. Given that $\tau \approx 1.618$, clearly $b_0 = 1$. Then $y_1 = \tau - 1 = -\tau' = 1/\tau$, so $x_1 = \tau$, which means that the process returns to exactly the same place. Thus, if one stops at any point, $\tau = [1; 1, 1, ..., \tau]$, or if one catties the process on forever, $\tau = [1; 1, 1, 1, ...]$ This is a particularly nice representation of the golden ratio τ.

The term *"convergent"* was used without definition above. It simply means the continued fraction obtained by stopping short of the end in a continued fraction expansion. So the convergents to τ are

$$[1], [1; 1], [1; 1, 1], [1; 1, 1, 1],$$

etc. And if one expands them into ordinary fractions, it is found that

$$[1] = 1 \ (= {}^{1}/_{1}),$$
$$[1; 1] = 2 \ (= {}^{2}/_{1}),$$
$$[1; 1, 1] = {}^{3}/_{2},$$
$$[1; 1, 1, 1] = {}^{5}/_{3},$$

and the successive *convergents* all take the form F_n/F_{n-1}. This is another way in which the golden ratio, the Fibonacci sequence, and continued fractions all relate.

It will be useful to derive a formula by which the convergents to a continued fraction can be calculated from the denominators. It is clear that

$$[b_0] = b_0,$$

by definition. And by expanding the fraction,

$$[b_0; b_1] = b_0 + 1/b_1 = (b_0 b_1 + 1)/b_1.$$

By the same reasoning,

$$b_1 + 1/b_2 = (b_1 b_2 + 1)/b_2, \text{ so that}$$
$$[b_0; b_1; b_2] = b_0 + 1/(b_1 + 1/b_2)$$
$$= b_0 + b_2/(b_1 b_2 + 1)$$
$$= [b_0(b_1 b_2 + 1) + b_2]/(b_1 b_2 + 1)$$
$$= (b_0 b_1 b_2 + b_0 + b_2)/(b_1 b_2 + 1)$$
$$= (b_0 b_1 b_2 + b_2 + b_0)/(b_1 b_2 + 1)$$
$$= [(b_0 b_1 + 1)b_2 + b_0]/(b_1 b_2 + 1).$$

If one writes $[b_0] = n_0/d_0$, $[b_0; b_1] = n_1/d_1$, $[b_0; b_1; b_2] = n_2/d_2$, etc., then

$$n_0 = b_0, \ d_0 = 1,$$

$$n_i = n_{i-1}b_i + n_{i-2}, \; d_i = d_{i-1}b_i + d_{i-2}.$$

(Here, formally, n_{-1} was set to 1 and d_{-1} to 0 where they appear in the expansion of n_1 and d_1.)

But, except for the fact that b_i is *not* a constant, but varies as i changes in the equations

$$n_i = n_{i-1}b_i + n_{i-2}; \; d_i = d_{i-1}b_i + d_{i-2},$$

this looks much like a **recursion** formula! And in fact, it looks like the recursion formula of two co-recurrent sequences! So the relationship of **continued** fractions to recurrent sequences is brought out here. And in fact, the numerators and denominators of the **convergents** to any continued fraction of the form [a; b, b, b, ...] (with a constant b) clearly form *two co-recurrent sequences*.

There is one more point that should be made at this place in the discussion. When the number that is being expressed in continued fraction form is *irrational*, the continued fraction does not terminate. But if there is a *large* number appearing as one of the denominators, stopping immediately before that point gives a particularly good rational approximation to the number in question. For example,

$$\pi = [3; 7, 15, 1, 292, 1, 1, ...]$$

$$[3] = 3,$$

$$[3; 7] = 3.142857...,$$

$$[3; 7; 15] = 3.14150943...,$$

$$[3; 7; 15; 1] = 3.14159292..., \text{ etc.}$$

The value reached by stopping at [3; 7] is the well-known $^{22}/_7$, which is particularly good for a fraction whose denominator is as small as 7. If you check, no other fraction with a denominator less than 100 comes anywhere near as close. And the last of these continued fractions, [3; 7; 15; 1], = $^{355}/_{113}$, which is accurate to 6 decimal places, amazingly good for a fraction whose denominator is as small as 3 digits! So, if the continued fraction for an irrational number has a large denominator, you can get a fraction particularly close to it by stopping right before that large number. In that sense, τ, whose continued fraction has only 1's for its denominators, is the *hardest* number to approximate by a fraction, which justifies the comment I have seen in some books that τ is the "most irrational number."

Chapter 37:   Recurrent sequences as aids to approximate computation of irrational numbers.

Because the terms of a recurrent sequence tend to become more like a geometric progression as the sequence progresses, with the ratio of a pair of consecutive terms approaching a limit, the use of recurrent sequences as a way of approximating irrational numbers suggests itself. For example, because $\tau = (1 + \sqrt{5})/2$, one can compute $\sqrt{5}$ in turn as $2\tau - 1$, and since $F_{n+1}/F_n \approx \tau$ for large n, one can use

$$\sqrt{5} \approx 2(F_{n+1}/F_n) - 1 \text{ (when } n \text{ is large)}$$

as a good approximation to τ (using the Fibonacci recurrence relation to compute F_n and F_{n+1}). In fact, since *any* generalized Fibonacci sequence has the same limiting ratio, one can speed the process by starting with G_0 and G_1 chosen to give a ratio that is already near τ and proceed from there. Since τ is known to be about 1.618, for example, one could use $^{1618}/_{1000} = {}^{809}/_{500}$ as a starting point, and construct a generalized Fibonacci sequence 500, 809, 1309, 2118, ..., with the following result:

n	G_n	G_{n+1}/G_n	Approximation to $\sqrt{5}$
0	500	1.61800000	2.23600000
1	809	1.61804697	2.23609394
2	1309	1.61802903	2.23605806
3	2118	1.61803588	2.23607177
4	3427	1.61803327	2.23606653
5	5545	1.61803427	2.23606853
6	8972	1.61803388	2.23606777
7	14517	1.61803403	2.23606806

Table 29: Approximation to $\sqrt{5}$ by ratios of terms of a generalized Fibonacci sequence.

This procedure is much easier and less error-prone than the normal way to calculate a square root, and thus it would seem recommended as a way to do it. The only problem is that one needs to find a recurrent sequence that will have a desired number as a limiting ratio, and of course, for any recurrent sequence that is not a generalized Fibonacci sequence, the recurrence relation requires some multiplications (by relatively easy multipliers, however) as well as the additions used here. What is more difficult is to figure out a particular choice of a recurrence relation that will yield the desired quantity. The key to this process is revealed in the fact that, for any recurrent sequence, if one writes the recurrence relation in terms of a given A_n and the preceding terms A_{n-1}, A_{n-2}, etc., and replaces A_{n-p} by x^p

wherever it occurs, the resulting polynomial-type equation has as its solutions the **common** ratios of the multiple geometric sequence that is equivalent to A, and the particular root that has the largest absolute value is the limiting value to which A_{n+1}/A_n tends for large n. This is why τ, which is a solution of the Fibonacci quadratic equation, is the limiting ratio of every generalized Fibonacci sequence. If one desires a solution to any polynomial-type equation, therefore, one can just use the equation, with x^p replaced by A_{n-p}, as a recurrence relation! But the problem is that a simple equation like $x^2 = a$ is awkward to use, because it has no term in x, giving a recursion formula which is not easy to apply.

It was noted, when the Binet formula for the Fibonacci sequence was stated, that the $\sqrt{5}$ terms in the expansion cancel out because they always appear in pairs with coefficients of opposite sign and equal magnitude. So it could be suspected that if a double geometric sequence were created with the two **common** ratios taking the form

$$r = p + q\sqrt{R},\ r' = p - q\sqrt{R},$$

the $\sqrt{R}$ terms would similarly cancel. And if one applies the double-geometric sequence to recurrent sequence formulas to these, the recursion coefficients obtained are

$$k_1 = r + r' = 2p,$$
$$k_2 = -rr' = q^2R - p^2.$$

And these, of course, are free of radical expressions, so that if p, q, and R are integers, so will k_1 and k_2 be. Now it is clear from what has been said about double geometric sequences earlier that if $|r| > |r'|$, the limiting ratio of A_{n+1}/A_n for large n will be equal to r (as long as one starts with a ratio A_1/A_0 which is different from r', no matter how slightly), and from that limiting ratio, therefore, a good approximation for $\sqrt{R}$ can be obtained.

It should be noted that the best choices for p and q would be such that $p \approx q\sqrt{R}$, making r' very close to 0; this is not the actual choice for the values obtained for the case of generalized Fibonacci sequences, though these best p and q values could, of course, have been computed. But in any case, just to demonstrate the process, let R = 7. Since $\sqrt{7} \approx 2.6$, a good choice for p and q would be $p = 3$ and $q = 1$, giving $k_1 = 6$ and $k_2 = -2$. Since this makes $A_{n+1}/A_n \approx 3 + \sqrt{7}$, one can approximate as $A_{n+1}/A_n - 3$. Table 30 demonstrates this process, in the same format as Table 29.

n	A_n	A_{n+1}/A_n	Approximate $\sqrt{7}$
0	1	1.00000000	-2.00000000
1	1	4.00000000	1.00000000

n	A_n	A_{n+1}/A_n	Approximate $\sqrt{7}$
2	4	5.50000000	2.50000000
3	22	5.63636364	2.63636364
4	124	5.64516129	2.64516129
5	700	5.64571429	2.64571429
6	3952	5.64574899	2.64574899
7	22312	5.64575117	2.64575117

Table 30: Approximation of a quadratic irrational by a recurrent sequence.

The exact value of $\sqrt{7}$ to 8 decimal places is 2.64575131, to show how close the result of this calculation is. It should be noted that the choice of $A_0 = A_1 = 1$ was totally arbitrary, and any two values could be used (except, as it happens, any choice that would make $A_1/A_0 = r'$). The only reason for this choice is that it makes the calculation simpler than just about any other choice would provide.

Chapter 38:   Bibliography.

Dunlap, Richard A.: *The Golden Ratio and Fibonacci Numbers.* Hackensack, N. J. (and other places): World Scientific Publishing Co., 1997. (This book and Vajda's below are the two best that I have seen on the subject.)

Knott, Ron: *Fibonacci Numbers and the Golden Section.* Web page at http://www.maths.surrey.ac.uk/hosted-sites/R.Knott/Fibonacci/ (Probably the best Internet site for Fibonacci-related material.)

Livio, Mario: *The Golden Ratio: The Story of PHI, the World's Most Astonishing Number,* 2002. New York, N. Y.: Broadway Books. (Although this author uses φ for the golden ratio, he makes it clear that the claims of its use by Phidias and others in the arts are unfounded. Clearly, he should take this argument to its logical conclusion and not use φ, but otherwise the book is rather good. As the title suggests, it is more about the golden ratio than about the Fibonacci sequence, but that they are intimately related makes its inclusion in this list appropriate.)

Posamentier, Alfred S.: *The Fabulous Fibonacci Numbers,* 2007. Amherst, N. Y.: Prometheus Books. (This book is in some ways very good, and in others not good at all. It is visually a more attractive book than Dunlap's or Vajda's, making good use of illustrations and tables to demonstrate some of the beautiful patterns associated with the Fibonacci sequence and the golden ratio (even though he insists on using φ for that number). It is *more popular* and somewhat *less mathematical* than Dunlap's or Vajda's book, and is thus an easier read. But unfortunately he uncritically repeats some myths that ought to be put to rest, such as Phidias' connection with the golden ratio and the theory that the Fibonacci sequence can be used to predict the stock market.)

Vajda, Steven: *Fibonacci and Lucas Numbers, and the Golden Section: Theory and Application.* Mineola, N. Y.: Dover Publications, 2008 (reprint of book originally published 1989). (See the comment on Dunlap's book, above.)

I do not know of a good book on continued fractions to recommend to readers at the level for which this book is designed. There are good chapters in both Dunlap's and Vajda's books, and some very useful material on Knott's web site, and I would probably recommend starting with those. If you are willing to handle more mathematics, consult a site like amazon.com or bookfinder.com for books with titles including the words "Continued fractions."

Index

www.ingramcontent.com/pod-product-compliance
Lightning Source LLC
Chambersburg PA
CBHW051533170526
45165CB00002B/713